Pioneers in Public Health

T0239642

The public health movement involved numerous individuals who made the case for change and put new practices into place. However despite a growing interest in how we understand history to inform current evidence-based practice, there is no book focusing on our progressive pioneers in public health and environmental health.

This book seeks to fill that gap. It examines carefully selected public and environmental health pioneers who made a real difference to the UK's health, some with international influence. Many of these pioneers were criticised in their life-times, yet they had the strength of character to know what they were doing was fundamentally right and persevered, often against many odds. The book includes chapters on:

- Thomas Fresh
- John Snow
- Duncan of Liverpool
- Margaret McMillan
- George Cadbury
- Christopher Addison
- Margery Spring Rice and others.

This book will help readers place pioneers in a wider context and to make more sense of their academic and practitioner work today; how evidence (and what was historically understood by it) underpins modern day practice; and how these visionary pioneers developed their ideas into practice, some not fully appreciated until after their own deaths. *Pioneers in Public Health* sets the tone for a renewed focus on research into evidence-based public and environmental health, which has become the subject of growing international interest in recent years.

Jill Stewart worked as an Environmental Health Officer specialising in private sector housing for several years before becoming a lecturer in London universities teaching housing, public health, environmental health and social work at undergraduate and post-graduate levels. She is a Corporate Member of the Chartered Institute of Environmental Health, a Fellow of both the Royal Society for the Promotion of Health and the Royal Geographical Society and an Associate Member of the Chartered Institute of Housing.

Routledge Focus on Environmental Health
Series Editor: Stephen Battersby, MBE PhD, FCIEH, FRSPH

Pioneers in Public Health: Lessons from History, edited by Jill Stewart

Pioneers in Public Health

Lessons from History

Edited by Jill Stewart

LONDON AND NEW YORK

First published 2017
by Routledge
2 Park Square, Milton Park, Abingdon, Oxon OX14 4RN

and by Routledge
711 Third Avenue, New York, NY 10017

Routledge is an imprint of the Taylor & Francis Group, an informa business

© 2017 selection and editorial matter, Jill Stewart; individual
chapters, the contributors

The right of Jill Stewart to be identified as the author of the
editorial material, and of the authors for their individual chapters,
has been asserted in accordance with sections 77 and 78 of the
Copyright, Designs and Patents Act 1988.

All rights reserved. No part of this book may be reprinted
or reproduced or utilised in any form or by any electronic,
mechanical, or other means, now known or hereafter invented,
including photocopying and recording, or in any information
storage or retrieval system, without permission in writing from the
publishers.

Trademark notice: Product or corporate names may be trademarks
or registered trademarks, and are used only for identification and
explanation without intent to infringe.

British Library Cataloguing-in-Publication Data
A catalogue record for this book is available from the British Library

Library of Congress Cataloging-in-Publication Data
Names: Stewart, Jill, 1967– editor.
Title: Pioneers in public health : lessons from history / edited by
   Jill Stewart. Other titles: Routledge focus on environmental
   health.
Description: Abingdon, Oxon ; New York, NY : Routledge, 2017. |
   Series: Routledge focus on environmental health | Includes
   bibliographical references and index.
Identifiers: LCCN 2017003391 (print) | LCCN 2017005703
   (ebook) | ISBN 9781138059450 (hbk. : alk. paper) | ISBN
   9781315163543 (ebk)
Subjects: | MESH: Health Personnel—history | Public Health—
   history | Environmental Health—history | England | Biography
Classification: LCC RA418 (print) | LCC RA418 (ebook) | NLM
   WZ 112 | DDC 362.1—dc23
LC record available at https://lccn.loc.gov/2017003391

ISBN: 978-1-138-05945-0 (hbk)
ISBN: 978-1-315-16354-3 (ebk)

Typeset in Times New Roman
by Apex CoVantage, LLC

For Bill Carey

Contents

Figures

Tables

Contributors

Dr Stephen Battersby is at Robens Centre for Public and Environmental Health, University of Surrey.

Matthew Clough is an Environmental Health Practitioner.

Dr Surindar Dhesi is in the School of Geography, Earth and Environmental Science, University of Birmingham.

William Hatchett is Editor, *Environmental Health News*.

Dr Susan Lammin is an Environmental Health Practitioner (retired).

Zena Lynch is in the School of Geography, Earth and Environmental Science, University of Birmingham.

Deirdre Mason is a freelance writer in public and environmental health.

Dr Alan Page is Associate Professor in Environmental Health, Middlesex University.

Dr Norman Parkinson is at King's College London, Department of Education and Professional Studies, School of Social Science and Public Policy.

Dr Jill Stewart is Senior Lecturer, Environmental Health and Housing, University of Middlesex.

Dr Hugh Thomas is Senior Lecturer in Public Health and Primary Care, St George's Medical School, London.

Ellis Turner is an Environmental Health Practitioner.

Series preface

This new series, Routledge Focus on Environmental Health, aims to explore environmental health topics in more detail than would be found in the usual environmental health texts. We also want to encourage readers to submit proposals so that we are responsive to the needs of environmental and public health practitioners and provide opportunities for first-time authors. This is a dynamic series, which aims to provide a forum for new ideas and debate on current environmental health topics. This is a new and exciting development for environmental and public health practitioners, particularly for new professionals. So if you have any ideas for monographs in the series please do not be afraid to submit them to me as series editor via the e-mail address below.

I have always encouraged new authors and for environmental health practitioners to "get published". All too often the work of EHPs goes unrecorded and unremarked and now with the demise of the *Journal of Environmental Health Research*, I am really pleased to have this opportunity to provide another route for practitioners to change this.

This initiative follows on from the publication of the 21st edition of *Clay's Handbook of Environmental Health* in July 2016, but that is largely a technical work and first point of reference. It is not intended that this series takes a wholly "technical" approach but will provide an opportunity to consider areas of practice in a different way, for example looking at the social and political aspects of environmental health in addition to a more discursive approach on specialist areas.

Also we recognise that "environmental health" can be taken to mean different things in different countries around the world. I know that *Clay's* no doubt has chapters that might not be relevant to some international practitioners; nevertheless, environmental health practitioners are part of the public health workforce in any country. So, this series will enable a wider range of practitioners and others with a professional interest to access information and also to write about issues relevant to them. Books in the series have a relatively short production time so that the information will also be more immediate than in a standard textbook or reference work.

We are working to have forthcoming monographs to cover such areas as Housing and Health, Air Pollution and Health, Potable Water, EHPs and Health and Wellbeing Boards, Ports and Borders and Public Health. That does not mean we have no need of further suggestions; quite the contrary, so I hope readers with ideas for a monograph will get in touch via *Ed.Needle@ tandf.co.uk*

Stephen Battersby MBE PhD, FCIEH, FRSPH
Series Editor

Preface

Public health pioneers: lessons from history

This first in the series of "Routledge Focus on Environmental Health" is also slightly different from what will be its successors in that it looks backwards rather than at current issues. However, the contributors have drawn lessons for current practise from their work. It explores the contribution of public health pioneers to the development of public and environmental health practise. It is sometimes helpful to pause and gain a historical perspective so that we have a better understanding of how problems have been tackled in the past and to learn lessons from our predecessors – and hopefully avoid repeating mistakes.

The pioneers are not all the most obvious candidates for a work such as this. There are several reasons for this. First the origin of this particular monograph was a proposed special edition of the *Journal of Environmental Health Research*, which is no longer published. The authors had been invited to write about a pioneer for whom they had high regard. Second, the more obvious and famous pioneers have been written about in many different places and probably need no fresh coverage. It is also important to learn from some of the less celebrated pioneers and their struggles to improve public and environmental health. All those who strive to improve public health stand on the shoulders of giants but there are more of these "giants" than we sometimes realise.

It is to be hoped that readers from the different strands of public health and all parts of the public health workforce will gain something from the contents.

Stephen Battersby MBE PhD FCIEH FRSPH

Acknowledgements

We are immensely grateful to the following individuals and organisations for allowing us to reproduce their material here:

Chapter 1: Introduction

The Chartered Institute of Environmental Health for permission to use the illustration of Edwin Chadwick

Chapter 3: John Simon

We appreciate guidance received from the Wellcome Library, part of Wellcome Trust, Euston, in respect of digitised images

Chapter 4: John Snow

Images courtesy of the Wellcome Library, part of Wellcome Trust, Euston

Chapter 10: Charles Booth

The London School of Economics and Political Science Library for permission to reproduce Booth's poverty map classification
The Wellcome Library, part of Wellcome Trust, Euston, for guidance on use of the illustration of Charles Booth

Chapter 13: Berthold Lubetkin

Jim Gritton for his photograph of Bevin Court, Islington

Other acknowledgements are cited where relevant in the body of the text.

1 Introduction

Jill Stewart

Chronologies of dates and events in history tell us only limited things of what happened in the past and can have little real meaning in our lives today. But history can be engaging; learning about not just what happened, but why it happened and what motivated people, can make history fascinating, lively and even fun. Bringing history, and its people, back to life makes it enticing and interesting through anecdotes, stories, pictures and film. It can shape what we think today.

This book tells stories of how some people in history, now long gone pioneers in public health, took forward their strong views, ideas and values and in so doing, took sometimes great personal risks which continue to make a difference to all of our lives today. They had little to work from; there is a sense that they had to start from (public health) scratch. Some here are selected as better-known and obvious pioneers whose names and dates will have featured in history books, but others will have not, and may have come under our authors' radars almost by chance, to now feature in this book.

What is immediately striking in reviewing many of these pioneers is their focus on what we would now call the Social Determinants of Health (Commission on Social Determinants of Health, 2008; Marmot et al., 2010) and these ideas and beliefs about how we could live healthier lives are far from new. In the modern day, we can reflect on what these pioneers archived and provided, and reinvigorate what they did in new forms. The social determinants of health are about conditions into which people are born, how they grow, work, live and age and how wider economic frameworks, politics and policies shape our development (Commission on Social Determinants of Health, 2008).

Others of our pioneers focused on mitigating the worst effects of polluted, contaminated environments and poor housing stock and sought to develop and deliver interventions that reduce risks to health. In fact, what emerges from many of the pioneers featured here is their contribution to the role of the contemporary environmental health practitioner, now one of the key professionals delivering on front line public health practice across a range of fronts with a focus in the social determinants of health as well as interventions to mitigate negative environmental health stressors (Burke et al., 2002).

What we can see from this is the ongoing relevance of and standing of the contemporary Environmental Health Practitioner (EHP), previously known as Public Health Inspectors, Sanitary Inspectors or earlier still as Inspector of Nuisance. Some notable researchers of environmental health who have focused on early EHPs (Inspector of Nuisance, later Sanitary Inspectors) and statutory nuisance continue to provide remedy conditions prejudicial to health.

Hamlin (2013) consolidates Victorian notions of nuisances into neighbours' complaints; early development of what should not be acceptable by community standards (rotten meat, leaky cesspools, noxious trades, overcrowded housing); inspection of public services; and finally, infected persons whose liberty of movement was seen to threaten everyone. Nuisance legislation was seen to help relieve a range of environmental threats and at a very local level. What seemed to be important was character rather than technical ability. It was not until later on that Inspectors of Nuisance were able to develop their own professional organisation. They remained subordinate to Medical Officers of Health; but were very busy with very explicit local bye-laws and statutes in creating local government to help "shape views of local government as competent and responsive" (Hamlin, 2013, p. 379).

Crook also comments on this gradual change to local government functions and the development of the accountable, interventionist and surveillance state by the end of the Edwardian period and for our purposes here, the role of the by then Sanitary Inspector. Derived from Chadwick-initiated reforms, roles still varied from place to place, but continued to develop and evolve as scientific understanding grew, most notably with germ theory, possibly – it is argued – at the "expense of traditional sanitary concerns" (Crook, 2007, p. 377), or even simply applying common sense; but the culture remained more reactive than proactive within a liberal culture of governance allowing for public rights and freedoms, alongside social order and health. Crook argues that this led to tensions between political or ethical tasks for the Sanitary Inspector and for him, the role is more flexible, nuanced and informed by a range of issues. In summary, Crook (2007, p. 393), reflecting on the interventionist and surveillance roles of Sanitary Inspectors, asks:

> Was it freedom from the tyranny of disease and discomfort, or freedom from the tyranny of government? Inspectors, naturally enough, were convinced it was the former.

Why bother with history?

What happened in the past and how can we understand and interpret events and developments in public health today? Different organisations take different perspectives in using history in contemporary policy drawing from heritage and tradition. Understanding what happened in history, and why, can help inform and enlighten current debate – and perhaps one key thing that is often forgotten, can make things far more interesting today. Historical events and interpretations can help us challenge our own viewpoints, our morals, ethics and beliefs, attitudes and inform more sensitive and effective interventions into the future. Policy would surely be weaker if we fail to understand where it has come from and why it shifted as it did.

History, heritage and the built environment provides a major contribution to contemporary public health (Berridge, 2008). History is also about power, ideology, attitude and value and many of our public health pioneers were born into positions of power; they had the right and perhaps felt an obligation, to use that power and make change.

The era, values and attitudes into which we were born can have a profound effect on what we believe, even if it is not founded in evidence or is outright wrong. Accepted beliefs around causation of disease and illness shape how we deal with it and our solutions are frequently founded in ideologies rather than evidence of what works. If for example we believe – without a shadow of doubt – that it is miasma, or foul air, that causes illness, it follows that what we need to do about it is related to that belief. It is therefore ironic that Edwin Chadwick, the father of the modern environmental health profession, was a miasmist, along with many others of his time. John Snow, conversely, was not; as a result he faced ridicule and it was not until after his death that those in a position of power accepted his evidence-based 'germ theory'.

Chadwick: miasma and environmental health

Those interested in public health will already be familiar with the life and times of Edwin Chadwick (1800–1890), who has been widely written about (for example Lewis, 1952), so we have not devoted a chapter to him in this book. Chadwick wanted people to understand that health was an issue for everyone with prevention being better than cure. He had been influential around matters of juvenile labour, factory inspectors and compensation for industrial employees. His long and wider-ranging career included appointment by the Royal Commission to examine Poor Laws and his reforms included Poor Law Unions, each with a workhouse and he later favoured new approaches to local administration. He enrolled at Law School in 1823, and became a barrister in 1830, when he made friends with philosophers including John Stuart Mill and Jeremy Bentham[1] and doctors including Thomas

Southwood Smith. His combination of circumstance and friendships led him to become increasingly interested in sanitary and health conditions reform and in 1842 he published 'The Sanitary Condition of the Labouring Population', researched and funded at his own expense (see Figure 1.1).

Chadwick of course pioneered the first Public Health Act of 1848, and believed that without compulsory legislation, there would be little progress. He called for reform to overcome the multiple complexities around the range of organisations and administrations involved in public health implementation. As Meneces (1972, p. 177) states: "slowly but with gathering momentum the new values which the 'Sanitary Idea' set before society were permeating the minds of law-makers and administrators."

The new legislation, related organisational and administrative reforms and the development of inter-professional relationships in the sphere of public health were to provide immense impetus for change. Chadwick had firmly established a relationship between the environment and health,

Figure 1.1 Edwin Chadwick

although he clung doggedly to his belief in miasma. By 1884 Chadwick was appointed as the first President of the Association of Public Sanitary Inspectors, now the Chartered Institute of Environmental Health. The head offices are named Chadwick Court.

Despite everything Chadwick achieved, he seems not to have made many friends, but perhaps that was the price of his progress and he has been described as obstinate and a "pest wherever he went" (Finer, 1951, p. 442). Meneces (1972, p. 352) reports: "that he was so hated was a distasteful fact which he faced unflinchingly, but, conscious of his own high purpose he was honestly bewildered that it was so."

Names of other pioneers crop up in reading about Chadwick, including the engineers Bazalgette (with whom he also had disagreements – Finer, 1951) and Rawlinson, concerning mainly sewers but also controls around common lodging houses and the Medical Officers of Health.

Beyond miasma: the unsung heroes

During the writing of this book, it became clear that there was a body of environmental health people interested in the history of their profession and in particular some of the pioneers who drove developments. These were people they considered their own local 'unsung heroes' and indeed there seems little – if anything – published on some of these pioneering souls. It also emerged that there are history societies and archives nationally with substantial untapped resources with a wealth of material about the history of environmental health and public health.

Bryan Boulter, a retired EHP from Portsmouth, forwarded details about a speech he was giving to a local history society about Sir Robert Rawlinson. Not surprisingly as a garrison town, Portsmouth faced huge public health challenges as the Public Health Act 1848 came into being. Mr Rawlinson was amongst the first inspectors to be appointed under the General Board of Health created by the new Act, a role that developed from his earlier position as borough surveyor and engineer. Public health problems facing him were overcrowding, unsatisfactory drainage and sanitation. Proposals were put in place to provide drainage, cheap pure water, by substituting water closets and soil pan apparatus for privies and cesspools and by maintaining good roads, courts, passages and footpaths. There were arguments about who should pay for what; particularly the contribution of the military. As a result, little progress was made in drainage until the 1860s. It was Robert Rawlinson who gained credit for using tact and skill to facilitate progress amongst the different factions.

Rawlinson also has national significance and headed a Sanitary Commission to remedy ill health and mortality during the Crimean War, travelling widely. He was then sent by Lord Palmerstone to Lancashire in 1863 following the collapse of the cotton industry to help support poverty-stricken communities; he provided for better water and drainage. He is said to have visited some 100 towns. In the 1860s he chaired a Royal Commission on the Pollution of Rivers and became Chief Engineer to the Local Government Board and sits alongside Bazalgette for his contribution to innovative sewerage systems. He was knighted in 1883 and was elected President of the Institution of Civil Engineers in 1894.

A further little known pioneer was Dr Thomas Shapter, brought to our attention by David Sexton, EHP, also of Exeter. Thomas Shapter graduated from Edinburgh and moved to Exeter in 1832, coincidentally as Asiatic cholera also arrived. He published his book *The History of the Cholera in Exeter 1832* (Shapter, 1849) and describes the arrival of cholera in the city, the residents' and authorities' reaction to it. The book sets the scene for environmental health conditions in the city and outlines symptoms, treatments, deaths and burial grounds as well as presenting detailed statistical data about cholera with comparisons across ages, duration, gender and relationship to other data; all basic epidemiological methods to find the casus and influences of cholera. His data indicated a population mortality rate of 1.42% and his data suggested a 1 in 3 chance of death once contracted, and that it affected the entire population similarly. His work was well received and cited by John Snow and whilst he did not link it to contaminated water, he had tried to understand its cause and observed that the intensity of cases, by drawing a map of the outbreak, were situated in crowded and ill-drained areas, which he referred to as 'social errors' and that he felt it was a preventable disease.

Back to the future

Compiling public health histories can be tricky; not all documents are available to us and archives may be few and far between and due to costs, non-digitised. This can lead to hours and days trawling through dusty papers before finding gems of information. It is there that our authors shine. Some have trawled through archives and in their chapters, acknowledged those individuals and organisations who have particularly helped them. Such localised, historical data varies from place to place, is hard to compare (if this is possible at all – see even Shapter and Snow) and there are also variations in geographical, social and economic contexts and we may use history in a very ad-hoc manner, but hopefully this book provides a useful starting point.

Each of our authors had been inspired to write about their pioneers from a different standpoint. For some, there was an obvious geographical connection, to their place of childhood or employment and a snowballing interest in the local significance of what an individual had achieved, be in a beautiful architectural contribution to the built environment, or a shift in policy and organisational thinking that was to have significant national – if not international – ramifications. For some of our authors, theirs was also an organic, personal journey of discovery and ongoing learning.

Research for other chapters also included searches using Google and Google Scholar, leading to wider identified reading, appraised literature from relevant academic journals, scrutiny of specific websites as well as reviews of literature and visual images to inform their writing. Many of our authors drew from the narrative in Hansard proceedings from the Houses of Commons and Lords, reviewing transcripts of speeches and debates as well as their pioneer's own published and unpublished papers, letters and archived materials and drew from their biographies. Many of these pioneers were themselves published writers, so resources were rich, exposing personal values and ideologies. Their ideas and drive features and survives them in Committee minutes, letters and personal papers.

Some of our authors here have published on aspects of social and political history before; for others it was a new endeavour and journey. Many also developed their thinking through visual appraisal techniques, considering for example the built environment (through Google Earth as well as photographs and illustrations).

Many archives exist, notably the digital archive maintained by the London School of Economics, London School of Hygiene and Tropical Medicine, Wellcome Foundation and St George's Medical School Archive and the authors are grateful for the time, support and help afforded to them. There is no doubt a mass of further archived material waiting to be explored. But there are also numerous other museums, places to visit and other archives, including the Chartered Institute of Environmental Health's library at Waterloo, also in South London. Beyond this, there are numerous examples of amazing buildings that have stood the test of time, designed as part of the wider and healthier built environment that we are still able to walk around and admire and continue to learn.

Perhaps it is because we are all inherently nosey; inspection and investigation are fundamental components of our day-to-day work in public and environmental health. We want to investigate and find out more about the evidence, theories and understand why people believed what they did, so much so that some faced ridicule and exclusion in their quest for understanding and their drive to change policy. Perhaps then it is also of little surprise that many of our authors here are part of the modern-day environmental health profession.

What this book is about

This book seeks to fill a gap in the current literature on these pioneers, some of whom are little known. At a time when change is more rapid than ever, this provides a contribution to our current thinking and a reminder of where we should be headed, and the lessons that can be learned. It is only really by appreciating history that we can more fully understand the complexities of health determinants and inequalities that shape contemporary health inequalities and inequities. Those referred to in this book challenged the status quo and demanded that things should – and could – be better. It provides a renewed focus on research into evidence-based public and environmental health, which has become the subject of growing international interest in recent years.

This book is not about what is happening now, but what shaped and informed what is happening now, and also why; how current practitioners would benefit from understanding more about how to focus on health determinants and outcomes and challenge when this does not take place based on evidence (often drawn from our history and influenced by ideology). This is a book to inspire today's public health workforce and the importance of learning from these pioneers.

As Berridge (2008) observes, history needs to be part of the process of evidence-based policy as it offers insights and interpretations about the past and her view is that without historical analysis, contemporary policy will be all the poorer.

Synopsis

Our pioneers are presented more or less in chronological order of their major contribution(s) to public health. As such chapters are ordered as follows:

Chapter 2 – Thomas Fresh: the first environmental health practitioner
Chapter 3 – Sir John Simon: a role model for public health practice?
Chapter 4 – John Snow: a pioneer in epidemiology
Chapter 5 – Sir Joseph Bazalgette: a man of persistence and vision
Chapter 6 – George Smith of Coalville ('the Children's Friend'): campaigner for factory and canal boats legislation
Chapter 7 – Duncan of Liverpool: the first Medical Officer of Health
Chapter 8 – Margaret McMillan: advocate and practitioner of improvements in children's health
Chapter 9 – George Cadbury and corporate social responsibility: working conditions, housing, education and food policy
Chapter 10 – Charles Booth's inquiry: poverty, poor housing and legacies for environmental health

Chapter 11 – Christopher Addison: health visionary, man of war, parliamentarian and practical pioneer

Chapter 12 – Margery Spring Rice: throwing light on hidden misery

Chapter 13 – Berthold Lubetkin: 'nothing is too good for ordinary people'

Conclusions will then be drawn in Chapter 14.

Note

1 From Stephen Battersby, Series Editor: Chadwick was a disciple of Bentham who instead of practicing the law decided to write about it, and he spent his life criticising the existing law and suggesting ways for its improvement. Although associated with the doctrine of Utilitarianism and the principle of 'the greatest happiness of the greatest number' – Bentham aimed to test the usefulness of existing institutions, practices and beliefs against an objective standard. He was an advocate of law reform, a critic of established political doctrines like natural law. He also had much to say on prison reform, religion, poor relief, international law and animal welfare. He advocated universal suffrage and the decriminalisation of homosexuality. UCL is his "home" and of course now is the "home" for Marmot and the IHE.

Bibliography

Battersby, S. (ed.) (2017) *Clay's Encyclopedia of Environmental Health* (21st ed.). Oxon: Routledge.

Berridge, V. (2008) History Matters? History's Role in Health Policy Making, Medical History, 52(3), 311–326. www.ncbi.nlm.nih.gov/pmc/articles/PMC2448976/ accessed 7 December 2016.

Burke, S., Gray, I., Paterson, K., and Meyrick, J. (2002) *Environmental Health 2012: A Key Partner in Delivering the Public Health Agenda*. London: Health Development Agency.

Commission on Social Determinants of Health. (CSDH) (2008) *Closing the Gap in a Generation: Health Equity Through Action on the Social Determinants of Health. Final Report of the Commission on Social Determinants of Health*. Geneva: World Health Organisation. http://apps.who.int/iris/bitstream/10665/43943/1/9789241563703_eng.pdf accessed 7 December 2016.

Crook, T. (2007) Sanitary Inspection and the Public Sphere in Late Victorian and Edwardian Britain: A Case Study in Liberal Governance. *Social History*, 32(4), 369–393, doi: 10.1080/03071020701616654

Finer, S. E. (1951) *The Life and Times of Sir Edwin Chadwick*. London: Methuen and Co. Ltd.

Hamlin, C. (2013) Nuisances and Community in Mid-Victorian England: The Attractions of Inspection. *Social History*, 38(3), 346–379, doi:10.1080/03071022.2013.817061

Hatchett, W., Spear, S., Stewart, J., Stewart, J., Greenwell, A., and Clapham, D. (2012) *The Stuff of Life: Public Health in Edwardian Britain*. London: Chartered Institute of Environmental Health.

Lewis, R. A. (1952) *Edwin Chadwick and the Public Health Movement 1832–1854*. London: Longmans, Green and Co.

Marmot, M. et al. (Institute of Health Equity) (2010) *Fair Society, Healthy Lives (The Marmot Review)*. Strategic Review of Health Inequalities in England Post 2010. www.instituteofhealthequity.org/projects/fair-society-healthy-lives-the-marmot-review accessed 7 December 2016.

Meneces, A. N. T. (1972) *Sir Edwin Chadwick: Public Health Pioneer* (a monograph published under the auspices of the Chadwick Trust).

Shapter, T. (1849) *The History of the cholera in Exeter in 1832*. London: John Churchill, Exeter: Adam Holden. https://archive.org/details/b21363821. This material has been provided by London School of Hygiene & Tropical Medicine Library & Archives Service. The original may be consulted at London School of Hygiene & Tropical Medicine Library & Archives Service.

Stewart, J. (2016) *Housing and Hope: The Influence of the Interwar Years in England*. iBook store; see also www.jillstewarthousing.co.uk

Stewart, J., Smith, D., and Gritton, J. (2016). A Public Health History of a Forgotten Corner of Kent: The Isle of Sheppey. Occasional Paper Number 3 available online from EHRNet, https://ukehrnet.wordpress.com/ehrnet-resources/occasional-papers/

Useful websites

Centre for History in Public Health, London School of Hygiene and Tropical Medicine
http://history.lshtm.ac.uk/
UK Environmental Health Research Network
https://ukehrnet.wordpress.com/

2 Thomas Fresh

The first environmental health practitioner

Norman Parkinson

Introduction

Today's 'environmental health practitioner' has a lineage that reaches back through the 'public health inspector' and 'sanitary inspector' to the 'inspector of nuisances' of the Victorian sanitary reform era (Parkinson, 2015). Thomas Fresh has often been cited as the first inspector of nuisances and environmental health practitioner (EHP) (e.g. Eastwood, 1998, p. 5; Handysides, 1999, p. 232; Macarthur, 2003, p. 68), though so far without a supporting rationale.

Thomas Fresh was born in 1803 in the Lake District village of Newbarns (now part of Barrow-in-Furness). He came from a wealthy and locally prominent family that had interests in property, farming, and the mining and founding of iron ore. Both his father and grandfather held local public offices. Fresh moved to Liverpool initially as an iron-founder, later being appointed a policeman. He travelled to the USA at least twice. Fresh was widowed twice and married three times. He left no children (Parkinson, 2013a, 2013b).

Fresh was one of the three pioneering officers appointed in January 1847 by the Borough of Liverpool under its private 1846 Sanatory Act (Great Britain, 1846), though I discovered that Fresh had held a non-statutory version of the post for several years. Liverpool was then Britain's second-largest town. The lives and work of the other two, Britain's first Medical Officer of Health, William Duncan, and Borough Engineer, James Newlands, have been well documented and celebrated. By contrast, until recently, little has been known about Thomas Fresh, Inspector of Nuisances (Eastwood, 1998; Yorke, 2009; Parkinson, 2013a). In 2014 the Chartered Institute of Environmental Health and the Formby Civic Society placed a blue plaque on Fresh's house, and an application is to be made to have the building 'listed' (Environmental Health News, 2015). Nevertheless, Fresh's true memorial lies in his professional achievements, the diverse nature of

the EHP's professional arena, and, surprisingly, in his founding of an afflu-
ent residential settlement that bears his name: Freshfield, between Formby
and Southport in Merseyside.

Legacy: Fresh's first appointments in Liverpool

Background

While Fresh may have been the first EHP, being the first inspector of nuisances appointed by a local authority health committee with an explicit public health role, he certainly was not the first inspector of nuisances, which
was an office of the ancient hundred, manorial and leet courts. The inspection
of nuisances was often the role of a (sometimes reluctant) nominated lay
member of the community or a lay 'leet jury', though larger jurisdictions
might employ full- or part-time inspectors of nuisances and some employed
rudimentary food inspectors known as 'leavelookers', 'bread and ale tasters',
'high tasters' and 'low tasters' (e.g. Birmingham Manorial Court (Stephens,
1964) and Manchester Manor Court Leet (Earwaker, 1884).

By the start of the sanitary reform era the powers of the medieval courts
were very much diminished. The lowest common tier of local government
was the parish. In some jurisdictions there was an inspector of nuisances; if
not, action against nuisances may have been taken by another parish officer,
perhaps the 'overseer', the constable or the beadle. Sometimes a 'jury' of
vestrymen would inspect and take action. Some parishes appointed inspectors of nuisances on rotation from the lay membership of the vestry: 'virtuous
but inexpert men' with but 'respectable intelligence' (Hanley, 2006, p. 724).

Some 250 large towns and cities were incorporated as Boroughs under
Royal Charters. Newcastle-under-Lyme was created a Borough in 1173 and
the earliest surviving archives of 1369 show that even then it employed food
inspectors or 'food wardens' (Briggs, 1973). By the start of the 17th century the Borough of Liverpool had a long-established food inspectorate of
'leavelookers' and 'alefounders', and it had been active in investigating and
controlling nuisances since at least the mid-16th century (Touzeau, 1910).

The reformation of the Boroughs started in 1833 with an investigative Select Committee and then a Royal Commission. The minutes of the
local inquiry into the Corporation of Liverpool confirm that there was a
'leavelooker', assisted by three 'market lookers' (Wilkinson and Hutton,
1833). The resultant 1835 Municipal Corporations Act (Great Britain, 1835)
created a new pattern of municipal boroughs elected annually by *ratepayers* rather than freemen. They were permitted to make bye-laws for the
good rule and government of their district and were required to establish
a watch committee and police force, but the legislation did not provide for

education, building regulation, drainage, sanitation, highways, street lighting, cemeteries, street paving, refuse collection or water.

From the late 18th to the mid 19th century, the provision of such services and infrastructure was met by private local Town Improvement Acts. These established an overlay of *ad hoc* local authorities known as 'Boards of Commissioners' responsible for a particular task: establishing a public water supply; sewering and paving streets; providing a refuse removal service; markets; street lighting; dealing with nuisances, etc. Under such a private Act, Bristol appointed a full-time Inspector of Nuisances in 1794 (Buer, 1926). In Manchester, Webb and Webb noted that the Commissioners in each ward were initially required personally to act as 'amateur inspectors of nuisances' (Webb and Webb, 1922), but Manchester certainly had two paid inspectors by 1824 (Baines, 1825).

Liverpool

While some towns and cities employed food inspectors and inspectors of nuisances long before the sanitary reform era, the two posts were distinct and the link to public health was not explicit. In Liverpool in the mid-19th century, the enforcement of the bye-laws and the common law of nuisance fell within the purview of the Watch Committee and its police. Under the agitation of William Duncan and others Liverpool promoted two private Acts: the Liverpool Building Act 1842 (Great Britain, 1842a) and the Liverpool Improvement Act 1842 (Great Britain, 1842b). Under the Improvement Act, the Borough established a 'Health of the Town Committee'. Initially this Committee had no officers of its own and so had to refer matters of public health concern to the Town Clerk for investigation and action. Thomas Fresh, by then a police inspector, was seconded to the Town Hall for such work, initially responding to references from the Watch Committee, but later taking on more and more tasks for the Health of the Town Committee (Parkinson, 2014). From the late 1830s Fresh gained a reputation for his 'intelligence, zeal and activity' (*Liverpool Mercury*, 1861).

The Liverpool Improvement Act also created a statutory post of 'Inspector of Slaughterhouses and Meat' whose role and powers extended to fish and food other than meat:

> (The Council shall appoint) . . . fit and proper persons as Inspectors of Slaughterhouses and Meat . . . authorized and empowered . . . to enter into and inspect any house or place . . . kept or used for slaughtering or killing any cattle of any kind . . . and in case . . . any cattle, carcase or part of a carcase shall be found and declared to be unsound or unwholesome or unfit for the food of man, the same shall be immediately burnt

or destroyed . . . he is hereby authorised to examine and inspect any meat hawked about for sale in any of the streets or exposed for sale in any other place . . . and if upon such examination and inspection any meat shall appear to be unsound or unwholesome or not fit for the food of man, it shall be lawful for such inspector to seize, take and carry away the meat.

(Great Britain, 1842b)

Fresh was appointed 'Inspector of Slaughterhouses and Meat' and Superintendent of Scavengers in addition to his other duties, thus bringing together general nuisance inspection, public cleansing and food inspection. The leavelookers were put under Fresh's control (Eastwood, 1998).

Contribution to contemporary policy and practice

The Liverpool Sanatory Act

The Borough's Minutes reveal that from 1842 Liverpool's progressive Health of the Town Committee increasingly began to pass matters to the Town Clerk and Fresh for action. The Minutes show that Inspector Fresh reported, among other things, on the state of a slaughterhouse; on nuisance from a chicory works; nuisance from the drains of an abattoir; the keeping of pigs; cellar dwellings; the smell from a manure works; smoke nuisance from an oil mill; and nuisance from a plaster of Paris works. When in June 1844, Fresh was briefly withdrawn from this work, the Health Committee objected strongly, and asked for him to be permanently and exclusively attached to them as their dedicated 'Inspector of Nuisances':

Inspector Fresh who discharged the Town Hall police duty with great ability ha(s) been withdrawn therefrom by the Head Constable, and the (Health) Committee (is) of the opinion that it is absolutely requisite that a suitable person belonging to the police force of rank, and with emoluments equal to an inspector's shall be appointed exclusively . . . which duties shall comprise the services of the police under this Committee.

(Borough of Liverpool, 1844a)

On 26 August the Committee formally requested his transfer:

Inspector Fresh should be permanently attached to the Town Hall duty, to act with respect to nuisances, licences . . . etc . . . considering the duties to be discharged by Inspector Fresh, his emoluments should be increased (from £90) to £100 per year besides clothing.

(Borough of Liverpool, 1844b)

Thus, because of his positive reputation, the historically precedential post was created for Fresh and on 4 September 1844, he became Liverpool's first public health officer and the first inspector of nuisances appointed by a UK local authority Health Committee (Borough of Liverpool, 1844c). It demonstrated an explicit shift in the emphasis of the role of the inspector of nuisances from dealing with 'unreasonable behaviour' to protecting public health. Action against nuisances without a public health connotation remained with the Watch Committee and their police.

Fresh's new title of 'Inspector of Nuisances' first appeared in the Minutes of the Health Committee of 25 November 1844 when complaints about premises used for the boiling of bones were referred to him for action (Borough of Liverpool, 1844d). The Minutes showed that during the remaining part of 1844 Fresh also reported on cellar dwellings, smoke nuisances and industrial smells. During 1845, he reported on nuisance from an abattoir, and on cellar and court dwellings, and he was given authority to advise a cleaning company on methods of cleansing courts and passages. In 1846, Fresh was asked by the Health Committee to report on objectionable cellars; on the bad state of slaughterhouses; and problems of nuisance from the manufacture of ammonium sulphate and the calcining of bones. He also reported to the Health Committee on cellar habitations; accumulations of filth and offensive matter; nuisance from middens and slaughterhouses; noxious effluvia from public graves; and nuisance from a chandlery. A review of the Minutes clearly shows that he was the principal council officer with responsibility for environmental health matters for years before the Liverpool Sanatory Act (Borough of Liverpool, 1842–1846).

Nationally, the concept of a local inspector of nuisances appointed to enforce sanitary regulation had been promoted by the House of Commons Select Committee on the Health of Towns and by the Health of Towns Association, and it was included in Lord Lincoln's 1845 Public Health Bill (Parkinson, 2014). When that bill failed, the Borough of Liverpool, motivated by its radical Health Committee, the Liverpool Health of the Town Association and the lectures of Dr William Duncan (Parkinson, 2014), promoted a private bill that became the precedential 1846 Liverpool Sanatory Act (Great Britain, 1846). The Act required the Borough to appoint a Medical Officer of Health (MoH) and an Inspector of Nuisances (IoN). Statutory status would provide continuity, and help to protect the officers from contrary interests within the Council and property interests in the Borough. The Borough may also have wished to pre-empt the imposition of national legislation.

No particular qualifications or person specification were mandated for the IoN. The Liverpool Act does not even qualify 'person' with 'suitable' or 'fit and proper'. Apart from the obvious fact that the role was new and no qualifications existed, this was probably because the job was seen as essentially one of law enforcement, and therefore the most appropriate candidate would be an experienced police officer. I also believe that it was because the Borough had already decided that the first appointee under its Act would be its existing inspector of nuisances, Thomas Fresh. Indeed, Fresh may well have been involved in the drafting of the bill.

Under the Act, the role of Liverpool's MoH was essentially 'horizontal': strategic epidemiological observation and medical advice; periodic reporting on the sanitary condition of the Borough; ascertaining the existence of disease and epidemics; 'pointing out' of those nuisances likely to cause and maintain diseases; 'pointing out' the means of preventing the spread of such diseases, and 'pointing out' the most efficient means of ventilation of various types of building. Initially, Duncan, the first MoH, had no staff of his own (e.g. Frazer, 1947; Chave, 1974) and sanitary operations were the responsibility of the Borough Engineer and the Inspector of Nuisances (IoN). The contemporary emphasis was on 'engineering' rather than 'medical' solutions to public health problems. Chadwick stated in 1842:

> The great preventatives are drainage, street and house cleaning by means of supplies of water and improved sewerage, and especially the introduction of cheaper and more efficient modes of removing all noxious refuse from the towns . . . the physician has done his work when he has pointed out the disease that results from the neglect of proper administrative measures, and has alleviated the sufferings of the victims.
>
> (Chadwick, 1842, p. 341)

An analysis of Council and Health Committee Minutes and the Liverpool Sanatory Act reveals that the statutory role of Liverpool's IoN had three main elements:

1   The receipt and investigation of complaints from the public of nuisances and contraventions of the Borough's bye-laws, rules and regulations, and the prosecution of related offenders;
2   The inspector of nuisances was designated the 'Inspector of Slaughter Houses and Meat' under the 1842 Liverpool Act. This was probably because both posts were already held by Fresh. It thus enshrined in statute the combination of general sanitary inspection with food surveillance; and
3   The management of street cleaning and the collection and disposal of solid waste – this included the collection, removal and disposal of the 'nightsoil' – the contents of privies and cesspits.

In his quadrennial report for 1847–1851, Fresh outlined the scope of the work of his department, that by then consisted of 13 technical and administrative staff (Fresh, 1851). The range of duties reflected contemporary concerns and scientific theories of the causes of disease, particularly, though not exclusively, miasmatic theories. Fresh's report suggests that he was certainly convinced of the contagious nature of some diseases, even if there was a discourse within medicine. Interestingly, there was no explicit reference to vermin, whether rodent or arthropod, for their role as vectors had not yet been established.

> The duties embraced in this department are the inspection and suppression of nuisances, the enforcement of the cleansing of the filthy and unhealthy dwellings in the low and crowded parts of the town; the superintendence of the arrangements necessary for the removal of middens; the regulation of cellar occupation; the adoption of proceedings for the prevention of the emission of smoke; the keeping of a slaughterhouse registry; the conducting of a registry of lodging houses, and the inspection of cemeteries, knackers' yards, etc.
>
> (Fresh, 1851)

Fresh added that these duties included dealing with overcrowded dwellings; dwellings without adequate water supply; leaking water supplies; animals kept as to be a nuisance prejudicial to health; accumulations and deposits; stagnant water; water and sewage percolating into cellars; unfenced excavations; various drainage and sewerage defects; black smoke from industrial premises and steam vessels; and houses and cellars after cases of infectious disease (Fresh, 1851).

Fresh did not inherit an existing department or a team of experienced, qualified personnel. He had to start from scratch, developing systems and procedures and selecting and training his own staff. His Nuisance Department and its systems became well known, and he received visits from other local authorities wishing to set up their own service (e.g. Borough of Liverpool, 1859b).

The Liverpool Act established a *direct* relationship with every inhabitant of the Borough, not just freemen or ratepayers, and not via the Council or the MoH. Every inhabitant had the right to make a complaint and there was an obligation on the inspector to independently investigate and, if appropriate, initiate legal proceedings. This direct relationship with the public, to listen to their complaints then investigate and attempt to resolve problems without fear or favour, became a central plank in the ideology of the environmental health practitioner. Furthermore, every inhabitant not only had a right to complain to the IoN, but also the right to 'inspect the books' to see what had been done about their complaint. This may be the UK's first example of 'open government'.

The Act did not prescribe the nature of the relationship between the IoN and the MoH. It created niches for two independent officers. Liverpool's inspector of nuisances reported directly to the Health Committee and the Council, not via the MoH. Subject to the provisions requiring the MoH's certificate, or reference to the local authority, Fresh enjoyed complete professional autonomy. In the National Archives I found that even after Duncan's appointment Fresh frequently wrote directly to the General Board of Health suggesting legislative improvements, and about epidemics in Liverpool (Fresh, 1849–1855). I also discovered that Fresh was a consultee to the government on draft public health legislation (Borough of Liverpool, 1851), and to the Privy Council on the future role of medical officers of health and their relationship with the inspector of nuisances (Brockington, 1965). Fresh was even invited to oversee sanitary arrangements in Ballaclava during the Crimean War (Soulby's Ulverston Advertiser, 1855; *Liverpool Mercury*, 1855).

Duncan's biographer, Frazer (1947), commented that the IoN was 'an officer independent of the Medical Officer of Health' yet 'obliged by the nature of his duties to work in close cooperation with him' (Frazer, 1947, p 41). The absence of a prescribed relationship laid the foundations for the difficulties between MoHs and inspectors that would exist for the next 125 years (e.g. Johnson, 1983). Yet, while it is clear that Duncan was uncomfortable with having no staff of his own (Frazer, 1947), I could find nothing in the Council minutes, Fresh and Duncan's quadrennial reports (Duncan, 1851) and Duncan's day books (Duncan, 1847–1859) that suggests any overt jurisdictional conflict.

The three pioneering officers seem to have worked well together; they have even been described as 'almost a cabal' (Ashton, 1997). Three years after they were appointed, a *Liverpool Mercury* editorial observed:

> Liverpool for a long time was notorious as being the most unhealthy town in the kingdom. The number of deaths in proportion to the population led to this conclusion . . . The deaths for the last month or two have been far below the average of former seasons, and there has been an astonishing decrease in those particular diseases which are generally considered to prevail in badly sewered and badly ventilated districts. Much good has undoubtedly been done by the Health Committee and their officers . . . A new map of Liverpool has been prepared, showing the contour lines, with the view of properly directing the levels for the drainage of the town; damp and unhealthy cellars have been cleared of their miserable occupants; courts have been flagged and paved, and in crowded and unhealthy districts they are periodically cleansed with pure water; court houses have been whitewashed and purified;

unhealthy trades in proximity to dwellings have been removed or the nuisance arising therefrom has been suppressed; lodging-houses, in which sometimes dozens of poor wretched creatures were indiscriminately huddled, have been licensed, the number of inmates limited, and, to prevent overcrowding, they are now regularly visited by officers of the police; owners of land have been compelled to rail and fence dangerous excavations, and to drain sheets of water from which putrid exhalations arose; slaughterhouses have been placed under strict supervision; delapidated and dangerous buildings have been removed, and cesspools, those pregnant sources of disease and death, are rarely now to be found . . . we have some of the most efficient officers.

(*Liverpool Mercury*, 1850)

Fresh's salary was then £170pa and Duncan's £750. The *Liverpool Mercury* commented:

either one is much overpaid or the other receives a very inadequate remuneration . . . we continually hear of the efficient services of Mr. Thomas Fresh, the Inspector of Nuisances, and his small but well directed staff of officers . . . The inspector, brings to his work a zeal and devotedness rarely witnessed in public service.

(*Liverpool Mercury*, 1850)

Fresh's success in the initial post created for him was reflected in the statutory post in Liverpool's Bill and subsequently mandated by its 1846 Sanatory Act. Thus Fresh's personal success had enduring national significance, for the Liverpool Act became the parliamentary precedent for other local legislation, including the City of London and the rest of the Metropolis, as well as national model legislation such as the Towns Improvement Clauses Act of 1847 and, to a lesser extent, the elective Public Health Act 1848, all of which mandated the appointment of an inspector of nuisances like Liverpool's.

From Fresh's background we know that like other sanitary inspectors of his era, he was 'a respectable man . . . presentable in private houses' (Richardson, 1892, p. 3). His job required 'a robust physique and undaunted character' (Maude,1953, p. 4), for the enforcement of public health regulations was not always popular, and an inspector was 'likely to be obstructed in his work and should at least be able to hold his own' (Vacher, 1892, p. 5). While it seems unlikely that Fresh went to university, it is clear that he was intelligent, literate, articulate and travelled. His background in the family businesses, and his life experience, would have enabled him to communicate with people from all walks of life: middle- and upper-class sanitary reformers and politicians; industrialists; property owners, slaughterhouse owners and butchers; as well as the labouring classes and people in extreme poverty. He had to bridge the gap between the terrible local environmental conditions suffered by the poor, and the middle-class politicians and professionals. Even well-intentioned sanitary reformers could not stomach the conditions that were everyday engagements to the inspector of nuisances. Edwin Chadwick wrote:

> My vacation has been absorbed with visiting with Mr Smith and Dr Playfair the worst parts of some of the worst towns. Dr Playfair has been knocked-up by it and has been seriously ill. Mr Smith has had dysentery; Sir Henry de la Beche was obliged at Bristol to stand up at the end of alleys and vomit while Dr Playfair was investigating overflowing privies.
>
> (Chadwick, 1843)

Despite Liverpool's officers being expressly forbidden from all private business activities (Borough of Liverpool, 1847), Fresh invested his inheritance in the development of 150 acres of land between Southport and Formby, Merseyside; an area since known as 'Freshfield'. He was also involved in the formation of a Building Society and, with other Council officers, in a mutual investment scheme. When in 1859 this became known to the Council, Fresh felt obliged to resign (Parkinson, 2013a, 2013b). Health Committee members expressed their deep regret at the loss of his services, and gave 'high compliment as regards the manner in which he had discharged his duties' (*Liverpool Mercury*, 1859a). The *Mercury* later affirmed that 'he had discharged his duties with integrity, and to the perfect satisfaction of his employers, for many years' (*Liverpool Mercury*, 1861) but, embarrassed, Fresh sold his Liverpool house (*Liverpool Mercury*, 1859b) and departed for the USA (*Glasgow Herald*, 1859).

On his return, one year later, he set up in business in Portrush, County Antrim, exporting to Liverpool's fish market (*Liverpool Mercury*, 1861). On 3 April 1861, in Glasgow, he was taken suddenly ill and died of pleurisy. He was interred at St James cemetery in Liverpool, close to the Anglican cathedral.

Conclusion

The environmental health practitioner evolved from the ancient offices of 'inspector of nuisances' and 'leavelooker'. Liverpool was the forerunner for the whole of the country when on 4 September 1844 Thomas Fresh, its inspector of nuisances, was transformed from a police inspector into an exclusive agent of the local authority's Health Committee. Fresh was responsible for meat and food inspection duties in addition to general sanitary inspection. The success of Fresh and his post was reflected in the Borough's promotion of a private Act of Parliament, and in January 1847 Fresh was appointed to the new statutory post of inspector of nuisances. The statutory post incorporated all of Fresh's existing duties. Fresh's role and the form and wording of Liverpool's Sanatory Act was used as a model in private legislation promoted by many other towns and in national statutes.

Fresh was intelligent, articulate, active and travelled. He developed public health administrative systems from scratch, and built up a model Nuisance Department. His private business activities eventually led to his departure from local authority service.

Note

For more biographical details of Thomas Fresh, see Parkinson, 2013a and 2013b.

Bibliography

Ashton, J. (1997) *Duncan of Liverpool*. Preston: Carnegie. Introduction to republished edition of Frazer, W. M. [1947].

Baines. (1825) Public Agents. *Baines Directory of Salford and Manchester*. Ancestry.com; *UK, City and Boundary Directories, 1600s–1900s*. http://search.ancestry. co.uk/content/viewerpf.asphx?h=905763&db=UKCityDirectories&indiv=try accessed 27 September 2011.

Borough of Liverpool (1842–1846) Data extracted from original *Minutes of The Health of the Town Committee* from 1842 to 1846, inclusive. Accessed at Liverpool Record Office, Liverpool City Libraries.

Borough of Liverpool (1844a) *Minutes of the Health of the Town Committee*. 3 June 1844. Accessed at Liverpool Record Office, Liverpool City Libraries.

Borough of Liverpool (1844b) *Minutes of the Health of the Town Committee*. 26 August 1844. Accessed at Liverpool Record Office, Liverpool City Libraries.

Borough of Liverpool (1844c) *Minutes of the Borough Council*. 4 September 1844. Accessed at Liverpool Record Office, Liverpool City Libraries.

Borough of Liverpool (1844d) *Minutes of the Health of the Town Committee*, 25 November 1844. Accessed at Liverpool Record Office, Liverpool City Libraries.

Borough of Liverpool (1847) *Minutes of the Health Committee, 12 July 1847*. Accessed at Liverpool Record Office, Liverpool City Library.

Borough of Liverpool (1851) *Minutes of the Health Committee, 17 April 1851*. Accessed at Liverpool Record Office, Liverpool City Library.

Borough of Liverpool (1859a) *Minutes of the Borough Council, 6 April 1859*. Accessed at Liverpool Record Office, Liverpool City Library.

Borough of Liverpool (1859b) *Minutes of the Health Committee, 7 April 1859*. Accessed at Liverpool Record Office, Liverpool City Library.

Briggs, J. (1973) *Newcastle Under Lyme, 1173–1973*. Newcastle-under-Lyme: Newcastle-under-Lyme Borough Council.

Brockington, C. F. (1965) *Public Health in the Nineteenth Century.* Edinburgh and London: E. and S. Livingstone.

Buer, M. C. (1926) *Health, Wealth and Population in the Early Days of the Industrial Revolution.* London: Routledge & Kegan Paul.

Chadwick, E. (1842) *Report on an Inquiry Into the Sanitary Condition of the Labouring Population of Great Britain.* Republished in Flinn, W. M. (ed.) (1965) *op cit.*

Chadwick, E. (1843) Visitations of the Commissioners: *Letter to Mr Graham, 7 December 1843.* Accessed at University College London, Edwin Chadwick Papers.

Chave, S. P. W. (1974) The Medical Officer of Health 1847–1974: The Formative Years. *Proceedings of the Royal Society of Medicine*, 67(12), 1243.

Duncan, W. H. (1847–1859) *Dr Duncan's Daybooks (flimsy copies of personal correspondence).* Accessed at Liverpool Record Office, Liverpool City Library. ref: 352.4 MOH.

Duncan, W. H. (1851) *Report to the Health Committee of the Borough of Liverpool on the Health of the Town 1847-1848-1849-1850, and on Other Matters Within His Department.* Liverpool: Harris and Co. Accessed at Liverpool Record Office, Liverpool City Library. ref: 352.4 MOH.

Earwaker, J. P. (1884) The Court Leet Records of the Manor of Manchester, from 1552, to . . . 1686, and from 1731 to . . . 1846. Manchester: H. Blacklock and Co. https://archive.org/details/courtleetrecord19coungoog. Accessed 3 February 2017.

Eastwood, M. (1998) Liverpool: a town ahead of its time. In: *For the Common Good: 150 Years of Public Health.* London: Chartered Institute of Environmental Health. Also personal communication of unpublished more extensive draft.

Environmental Health News. (2015) CIEH Bulletin, *Environmental Health News*, December 2014/January 2015, 10.

Flinn, W. M. (ed.) (1965) *Chadwick's 'The Sanitary Condition of the Labouring Population of Great Britain 1842'.* Edinburgh: Edinburgh University Press.

Frazer, W. M. [1947] (1997) *Duncan of Liverpool.* Preston: Carnegie.

Fresh, T. (1849–1855) Letters to the General Board of Health. Files refs: MH/115/10, MH/115/16, MH/115/17, MH/115/35, MH/115/39, MH/115/40, MH/115/80, MH/115/93. Accessed at the National Archive, Kew.

Fresh, T. (1851) *Sanitory Operations in the Nuisance Department 1/1/1847 to 31/3/1851.* Liverpool: Harris & Company. Accessed at Liverpool Record Office, Liverpool City Library, ref: 352.4 NUI.

Glasgow Herald (1859) *Glasgow Herald* 12 May 1859; issue 6130.

Great Britain (1835) *Municipal Corporations Act 1835.* 5 & 6 Will. Ch. 76. accessed at Senate House Library Special Collections.

Great Britain (1842a) *The Liverpool Building Act of 1842.* accessed at Liverpool Record Office, Liverpool City Library.

Great Britain (1842b) *The Liverpool Improvement Act of 1842.* accessed at Liverpool Record Office, Liverpool City Library.

Great Britain (1846a) *The Liverpool Sanatory Act, 1846.* 9 & 10 Vict. Ch. 127. London: George E. Eyre and William Spottiswoode. Accessed at Senate House Library Special Collections.

Handysides, S. (1999) All the History You Can Remember. *Communicable Disease and Public Health*, 2(4), 232.

Hanley, J. (2006) Parliament, Physicians and Nuisances. *Bulletin History of Medicine*, 80, 724.

Johnson, R. (1983) *A Century of Progress, the History of the Institution of Environmental Health Officers, 1883–1983*. London: Institution of Environmental Health Officers.

Liverpool Mercury (1850) Sanatory Reform in Liverpool. 14 May 1850; Issue 2193.

Liverpool Mercury (1855) The Sanitary Commission for the East. *Liverpool Mercury, 23 February 1855; Issue 2666*.

Liverpool Mercury (1859a) Health Committee. 8 April 1859; Issue 3477.

Liverpool Mercury (1859b) Sales by Auction, Classified Advertisements. 18 March 1859; Issue 3459.

Liverpool Mercury (1861) Obituary of Thomas Fresh. 6th April 1861; Issue 4103.

Macarthur, I. (2003, March) Making a World of Difference. *Environmental Health Journal*, 111(3), 68–70.

Maude, J. (1953) *Report of the Working Party on the Recruitment, Training and Qualifications of Sanitary Inspectors*. London: Ministry of Health, HMSO.

Parkinson, N. (2013a) Thomas Fresh (1803–1861), Inspector of Nuisances, Liverpool's First Public Health Officer. *Journal of Medical Biography*, 21(4), 238–249.

Parkinson, N. (2013b) Thomas Fresh 1803–1861. Paper presented to the Local History Group, Formby Civic Society.

Parkinson, N. (2014) The Health of Towns Association and the Genesis of the Environmental Health Practitioner. *Journal of Environmental Health Research*, 14(1), 5–16.

Parkinson, N. (2015) 'Inspector of Nuisances' to 'Environmental Health Practitioner': A Case Study of Title Change in the Professionalisation Process. *Journal of Environmental Health Research*, 15(1), 4–22.

Richardson, B. W. (1892) quoted in Johnson, R., op. cit.

Soulby's Ulverston Advertiser (1855, February 22) Inspector Fresh of Liverpool and the Government. *Soulby's Ulverston Advertiser*.

Stephens, W. B. (1964) Political and administrative history: Local government and public services. In: *A History of the County of Warwick: Volume 7 The City of Birmingham*, pp. 318–353. www.british-history.ac.uk/report.aspx?compid=22973 accessed 7 October 2010.

Touzeau, J. (1910) *The Rise and Progress of Liverpool From 1551–1835*. Accessed at Liverpool Record Office, Liverpool City Library, ref: 942.721 TOU.

Vacher, F. (1892) *The Food Inspector's Handbook*. London: Sanitary Publishing Company.

Webb, S., and Webb, B. (1922) *English Local Government: Statutory Authorities for Special Purposes* (Vol. 4). London: Longmans Green.

Wilkinson and Hutton (no initials) (1833) *A Report of the Proceedings of a Court of Inquiry into the Existing State of the City of Liverpool*. London: H. M. Government; Liverpool: J. & J. Mawdesley. Accessed at Liverpool Record Office, Liverpool City Library, ref: 352.09 INQ.

Yorke, R. (2009) Thomas Fresh: Inspector of Nuisances. *Journal of the Liverpool History Society*, 8, 16–24.

3 Sir John Simon

A role model for public health practice?

Alan Page

Introduction

Sir Donald Acheson, Sir Kenneth Calman, Sir Liam Donaldson and Professor Dame Sally Davies are names familiar to many reading this, as recent and current Chief Medical Officers of Her Majesty's Government. Their forebear though is probably less familiar to us. Sir John Simon (1816–1904), pronounced in the French form, was the City of London's first Medical Officer of Health and the first Chief Medical Officer for England and Wales. Born of French parents, he was brought up in the City of London and was apprenticed as a surgeon at the age of 17 and admitted to the Royal College of Surgeons in 1838 (Lambert, 1963).

Before exploring Sir John Simon's role as a public health pioneer, it is important to set a context in which he worked. Unlike the milieu in which environmental health and public health practitioners now work, mid-19th-century England had no framework for public health, no legislation to tackle the abundant public health issues, nor a clear understanding of the propagation of communicable disease, or a surveillance system to monitor such disease, with the exception in the latter case of cause of death contained in the Bills of Mortality (Szreter, 1988). This is a period of endemic smallpox (Hardy, 1993), periodic cholera epidemics, and where the average age of death in the 1830s and 1840s was around 29 in urban areas (Twiss, 1845; Daunton, 2004).

It is also a period of massive migration and economic growth. In a marked resemblance to the modern world, migrants experienced poorer health due to "residential segregation and low socio-economic status" (Davenport, Boulton and Schwartz, 2010). As Szreter (1997, 2001) argues rapid economic growth results in critical social insecurities, change and health issues resultant from disruption. That the overall economy prospers, it is at the expense of the poor and unless there is a rapid state and political response this disruption leads to deprivation, disease and death (Szreter, 1999). This migration resulted from urban pull and the increasing population size in the UK resulting from increased fertility (Wrigley and Schofield, 1983; Wrigley, 1985), a result of earlier marriage and reduced mortality which had been evident throughout the 18th century. However the period 1830–1870 saw this reduced mortality stagnate through evidenced decline in sanitary conditions.

The high incidence of communicable disease and net migration required systematic changes in public health behaviours and sanitation (Armstrong, 1993). As Armstrong (1993) highlights the period leading up to the reforms instigated by the EH pioneers was dominated by "cordon sanitaire" in which quarantine was utilised to segregate the sick from the healthy using spatial separation which often remained even once the sick moved on, died or recovered.

Alongside the above it has been noted that this was also a period in which liberal politics were a dominant force. Legislative interventions such as the Vaccination Act (1853) and the Contagious Diseases Act 1864 (Porter and Porter, 1988) provoked huge controversy through the compulsion of vaccination, something that has returned to trouble many in society today (Kennedy, Brown and Gust, 2005; Bean, 2011). The term "liberal" though is used with some caution. It is not used in the context of non-state intervention or laissez faire but instead used in the context of advocation of progressive reform and as Crook (2007) argues the desire for a moralised and health society capable of self-governance through guidance and rules.

Legacy

It is into the parlous public health context that Sir John Simon, a renowned surgeon, researcher and the first lecturer in pathology in the UK (Lambert, 1963) journeyed. Coming from a "socially prominent family" and embedded as a surgeon, Simon was seen as uncontroversial an appointment to the role of Medical Officer of Health for the City of London (UCLA undated) and indeed a committee was formed to support his appointment (Lambert, 1963). To this role, however, he applied his skills as a scientist insisting on having evidence before suggesting reasons or solutions. His first annual report of 1849 set the tone of his subsequent work, embedded with detailed analysis of mortality rates and evaluation of the causation. A staunch advocate of the links between poor housing and poor health he advocated slum clearance, building of hygienic (model) housing, control of water supplies including the provision of public wash houses, controls over burials and restrictions over offensive trades such as bone boiling, leather making, animal hide sales (Crook, 2007; see Table 3.1).

Table 3.1 Report on the sanitary condition of the City of London for the year 1848–9, p. 1

Mortality of the City of London	3
House-Drainage (sic)	12
Water-Supply (sic)	22
Offensive and Injurious Trades	29
Intramural Burial	36
Houses Permanently Unfit for Habitation	44
Social Condition of the Poor	54
Suggestions for Sanitary Organization in the City	70
Recapitulation and Conclusion	79
Tables	86

Source: Simon (1849).

This report was seen as controversial, with the aldermen of the City accusing him of gross exaggeration (Lambert, 1963). However the facts were clear, the City of London had a mortality rate of 30 deaths per 1,000 population, a figure as Simon (1849) highlighted was nearly three times the worst suburban district rate of 11 deaths per 1,000 population evidenced by the Registrar General's Reports of the same period. At that time it was all too easy to blame epidemics of cholera for high mortality levels; however, Simon turned his attention not to the presence of cholera but to the conditions that lead to its incidence and prevalence and arguably to a movement towards preventative interventions (see Figure 3.1).

Likewise he explored, in great detail, the differing mortality rates within the districts of the City of London to highlight that the aggregated high mortality rate for the City was skewed significantly by a number of districts and thus action should be focused rather than generally applied. See Figure 3.2.

The foundation of his argument may not always have been correct, relying on miasma as the causative agent, "These gases, which so many thousands of persons are daily inhaling, . . . rise from so many cesspools, and taint the atmosphere of so many houses, they form a climate the most congenial for the multiplication of epidemic disorders" (Simon, 1849 p. 13), but his recognition that by removing the source of harm public health could be improved was a significant shift in thought. At that time many in society saw poverty as a result of indolence and thus the poor were deserving of their predicament. These were the advocates of the workhouse in which conditions were deliberately poor so as to discourage dependency on the state or parish (de Pennington, 2011). The others such as Simon were social reformers, suggesting that it was circumstances beyond their control that led the poor to poor health. That this debate remains current (Chapman, 2012) and that it is a former Archbishop of Canterbury that was embroiled

Such is the general rule; and accordingly I would suggest to you that the presence of epidemic cholera, instead of serving to explain away the local inequalities of mortality, does, in fact, only constitute a most important additional testimony to the salubrity or insalubrity of a district, and renders more evident a disparity of circumstances which was previously decided. The frightful phenomenon of a periodic pestilence belongs only to defective sanitary arrangements . . .

Figure 3.1 Report on the sanitary condition of the City of London for the year 1848–9, p. 7

Source: Simon (1849).

These facts are quite unquestionable, and I have felt it my duty to bring them under your notice as pointedly and impressively as I can; feeling assured, as I do, that so soon as you are cognizant of them, every motive of humanity, no less of economical prudence, must engage you to investigate with me, whether or not there may lie within your reach any adoptable measures for lessening this large expenditure of human life, and for or relieving its attendant misery. It is, therefore, with the deepest feeling of responsibility that I proceed to fulfil the main object of my First Annual Report, by tracing these effects to their causes, and by explaining to you, from a year's observation and experience, what seem to me the chief influences prevailing against life within the City of London.

Figure 3.2 Report on the sanitary condition of the City of London for the year 1848–9, p. 10

Source: Simon (1849).

in the debate shows how Simon stood out in his advocacy for sanitary improvement.

The report also highlighted the importance of water supply, both in terms of quality but also quantity. As his reports highlights:

that its unrestricted supply is the first essential of decency, of comfort, and of health; that no civilization of the poorer classes can exist without it; and that any limitation to its use in the metropolis is a barrier, which must maintain thousands in a state of the most unwholesome filth and degradation.

(Simon, 1849, p. 22)

At this time a communal well, cistern or butt would be common and even where water was supplied to houses directly "the water is turned on only for a few hours in a week" (Simon, 1849, p. 23) requiring storage of water for prolonged periods subjecting it to potential contamination "receiving soot and all other impurities from the air, absorbing stench from the adjacent cesspool; inviting filth from insects, vermin, sparrows, cats and children . . . and every hour becoming fustier and more offensive" (Simon, 1849, p. 23).

His recommendations were bold suggesting potable water to all households and not just houses, thus including tenements and lodging houses, the forebear of the modern House in Multiple Occupation. He also suggested that privies should be directly supplied with water to enable flushing and cleansing of the discharge pipe. On offensive trades he called for all slaughtering of animals in the City of London to be banned. At the time there were 138 slaughterhouses of which 58 were in "vaults and cellars" (Simon, 1849, p. 32). He was also a pioneer in managing "nuisance" trades such as tallow melting suggesting that it is a common right to breathe an uncontaminated atmosphere, again something that has returned to haunt those living in London. His use of terms common to those within environmental health in relation to such nuisances underpin today's practice, for example,

It might be an infraction of personal liberty, to interfere with a proprietor's right to make offensive smells within the limits of his own tenement, and for his own separate inhalation; but surely it is a still greater infraction of personal liberty when the proprietor, entitled as he is to but the joint use of an atmosphere, which is the common property of his neighbourhood.

(Simon, 1849, p. 34)

Likewise he focused attention on smoke nuisance both detrimental to health and wellbeing but also from the perspective of "wasted fuel" thus linking both environmental degradation and health.

His 1850 report is more politically crafted. The data set on which he reports showed a notable reduction in the mortality rate, which placed his advocacy of future intervention on difficult ground. He applauds the work of the committee to whom he reported but continues his activism for the removal of "evils" such as overcrowding, defective drainage, contamination of water supplies pointing to the limited data sets over the two years of his incumbency. There are marked differences in the report style from his 1849 report, becoming distinctly more evidence-based (see Table 3.2).

Table 3.2 No. V. Comparative prevalence during the three years 1848–9, 1849–50 and 1850–1 of such deaths from acute disease as may chiefly be considered preventable

In the Years terminating as follows: –	Fever &c.	Cholera, Dysentery Epidemic Diarrhoea	Scarlet Fever Cynanche Maligna, &c.	Small Pox, &c.	Erysipelas Puerperal Fever, Phlebitis, &c.	Diarrhoea, Bronchitis and Pneumonia of infants, under 3 years of age	Zymotic Disorders commonly occurring in Infants: Hooping-cough, Croup, Measles	Hydrocephalus and Convulsions of Infancy	Total.
At Michaelmas 1849	166	825	135	17	44	285	196	264	1,932
,, 1850	118	54	32	33	40	243	124	219	863
,, 1851	107	23	46	41	17	340	272	282	1,128
Total of three years	391	902	213	91	101	868	592	765	3,923

Source: Simon (1850).

By 1853 his data sets were able to demonstrate some form of trend, although as he rightly points to, the absence of data prior to the interventions made for difficult interpretation of efficacy or effect. This is something that modern environmental health practitioners may well reflect upon, as without a base line datum it is neigh on impossible to assess the effectiveness of any given intervention. That he saw this some 160 years ago shows the enlightened worldview that he offered public health. What he also began to extol was the notion of multiple causality exploring urban planning, drainage, population density, relative height of occupied ground compared to the nearest water surface (Simon, 1853, p. 20).

As Lambert (1963) highlights he faced considerable opposition to his interventions but was aided by the ever-present cholera epidemic which led to an increasing mortality rate which Simon used to demonstrate the necessity of change.

Implications for contemporary policy and practice

His work was authoritative, evidenced and detailed and was recognised by prominent medical practitioners such as Charles Witt who in a letter to Simon in 1859 records his "humble tribute of thanks for your unwearied labours as a guardian of public health" (Witt, 1859, p. 1). There are lessons for environmental health and wider public health practitioners that practice must be evidence-based but of equal importance is the need for the voice of the practitioner to be heard and acted upon. The former is something that practitioners can act upon, but the latter requires an additional skill set of political awareness and in this case tenacity to seek sustained change.

Such was his effect that he then was promoted to medical officer for the General Board of Health (GBH), the first national role, showing at that time that his voice indeed was being heard and the efficacy of his actions recognised. That said the role did not last long as the GBH was abolished, a point at which many practitioners navigating the complexity of power would falter. Instead Simon's role was transferred to the Privy Council, a position of substantial power, which enabled him to ensure his views were embodied in legislation. Seymour (2007) points to the more contested and complex arena in which health policy is now framed between the Chief Medical Officer, Permanent Secretary and Minister and where the tensions between meeting the needs of government, the nation's health and the medical profession are highly charged.

Conclusions

In a remarkable rehearsal of the modern era Simon was constantly challenged by arguments surrounding central government and localism with local authority advocates such as Chadwick strong opponents. Likewise whilst managing to enshrine his public health message within legislation, on occasions his interventions were undermined by changes in political position with legislation watered down by subsequent administrations (Lambert, 1963) and by 1871 he had lost his access to Parliament becoming an advisor rather than an executive agent of government. Finally again in a remarkable echo of the future on his retirement the appointment within the Privy Council was abolished and the role of medical officer to the Local Government Board downgraded in both salary and prestige, much as has happened within public health today.

Bibliography

Armstrong, D. (1993) Public Health Spaces and the Fabrication of Identity. *Sociology*, 27(3), 393–410.

Bean, S. J. (2011) Emerging and Continuing Trends in Vaccine Opposition Website Content. *Vaccine*, 29(10), 1874–1880.

Chapman, J. (2012, January 25) Well, hallelujah! Archbishop Blasts Clerics Who Oppose Welfare Reforms and Declares the REAL Moral Scandal Is Our £1TRILLION Debt. *Daily Mail*.

Crook, T. (2007) Sanitary Inspection and the Public Sphere in Late Victoria and Edwardian Britain: A Case Study in Liberal Governance. *Social History*, 32(4), 369–393.

Daunton, M. (2004) *London's Great Stink and Victorian Urban Planning, BBC History Trails: Victorian Britain.* www.bbc.co.uk/history/trail/victorian_britain/social_conditions/victorian_urban_planning_01.shtml accessed 11 March 2016.

Davenport, R., Boulton, J., and Schwartz, L. (2010) *Infant and Young Adult Mortality in London's West End, 1750–1824.* Working paper Newcastle University: pauper lives project. http://research.ncl.ac.uk/pauperlives/infantandchildmortality.pdf accessed 11 March 2016.

Hardy, A. (1993) *The Medical Response to Epidemic Disease During the Long Eighteenth Century in Champion J.A.I (1993) Epidemic Disease in London.* Centre for Metropolitan History Working Papers no 1. Centre of Metropolitan History.

Kennedy, A. M., Brown, C. J., and Gust, D. A. (2005) Vaccine Beliefs of Parents Who Oppose Compulsory Vaccination. *Public Health Reports*, 120(3), 252–258.

Lambert, R. (1963) *Sir John Simon 1816–1904 and English Social Administration.* London: Macgibbon and Kee. www.cabdirect.org/abstracts/19642700821.html;jsessionid=F03AE1A85FE0DE21B4C83F8AAAED2489?freeview=true accessed 11 March 2016.

Pennington, J. de. (2011, February 17) Beneath the Surface: A Country of Two Nations. *BBC History.* www.bbc.co.uk/history/british/victorians/bsurface_01.shtml accessed 16 June 2016.

Porter, D., and Porter, R. (1988) The Politics of Prevention: Anti-Vaccinationism and Public Health in the Nineteenth Century. *Medical History*, 32, 231–252.

Seymour, J. K. (2007) Book Review: The Nation's Doctor: the Role of the Chief Medical Officer 1855–1998. *Medical History*, 51(3), 427–428.

Simon. (1849) Report on the sanitary condition of the City of London for the year 1848–9.

Simon. (1850) Report on the sanitary condition of the City of London for the year 1849–50.

Simon. (1853) Report on the sanitary condition of the City of London for the year 1852–53.

Szreter, S. (1988) The Importance of Social Intervention in Britain's Mortality Decline c 1850 to 1914: A Re-Interpretation of the Role of Public Health. *Social History of Medicine*, 1, 1–37.

Szreter, S. (1997) Economic Growth, Disruption, Deprivation, Disease and Death: On the Importance of Politics of Public Health for Development. *Population and Development Review*, 23(4), 693–728.

Szreter, S. (1999) Rapid Economic Growth and "the four Ds" of Disruption, Deprivation, Disease and Death: Public Health Lessons From Nineteenth-Century Britain for Twenty-First Century China? *Tropical Medicine and International Health*, 4(2), 146–152.

Szreter, S. (2001) *Economic Growth, Disruption, Deprivation, Disease and Death: On the Importance of the Politics of Public Health for Development in A.T. Price-Smith Plagues and Politics*. London: Palgrave Macmillan.

Twiss, T. (1845) *On Certain Tests of a Thriving Population: Four Lectures Delivered Before the University of Oxford in Lent Term*. https://play.google.com/books/reader?id=shkEAAAAQAAJ&printsec=frontcover&output=reader&hl=en_GB&pg=GBS.PA34 accessed 17 June 2016.

UCLA. (undated). *UCLA Epidemiology*. John Snow site: London historical references and sights: Sir John Simon. www.ph.ucla.edu/epi/snow/1859map/simon_john.html accessed 17 June 2016.

Witt, C. (1859) *Letter to John Simon F.R.S.* Wellcome Library: http://wellcomelibrary.org/item/b22434963#?m=0&cv=8&h=JOHN+SIMON&c=0&s=0&z=-1.2573%2C-0.0963%2C3.5148%2C1.9268 accessed 1 July 2016.

Wrigley, E. A. (1985) Population Growth: England, 1680–1820. *ReFRESH 1 Autumn 1985*.

Wrigley, E. A., and Schofield, R. S. (1983) English Population From Family Reconstitution: Summary Results 1600–1799. *Population Studies*, 37(2), 157–184.

4 John Snow

A pioneer in epidemiology

Hugh Thomas

Introduction

In the 19th century Britain experienced four cholera epidemics, each resulting in thousands of deaths. They were of a virulent form of 'Asian cholera' with a pattern of spread that seemed to show origin in India, spread through Russia and Europe and probable entry to England via the Hanseatic ports to our large ports including Tyneside and London. The British government used its diplomatic service to try to monitor the disease's spread but, like the medical profession, had no idea what caused the disease, or how it could be prevented or treated. The Russians had tried quarantine but this was hard to enforce and ineffective. Numerous unproven theories existed about the cause and numerous treatments were tried unsuccessfully. The disease was sometimes called the 'Blue Disease' because it was noted that cases often had blueish distorted fingers and toes, widespread cramps, severe vomiting and diarrhoea with 'rice water' stools and corrugated dry skin. Death often occurred within a few days of falling ill. See Figure 4.1.

John Snow born in York in 1813, the son of a farmer, and it is generally thought that it was a benevolent family friend who financed him to take on an apprenticeship. He was an unusual apprentice being a founder member of the York Temperance Society and also a vegetarian. He started his training at the age of 14 years and first encountered cholera in 1832 as a 19-year-old during his training in Newcastle. Snow had begun to make observations on the disease and its sufferers and noted that, unlike many other plagues, the physicians trying to treat it were not often its victims. He also did not think that cholera was caused by bad air or dirt as people working in offensive trades, often associated with 'nuisances' and 'noxious effluvia', such as boiling bones, removing 'night soil' or metal grinding, did not have higher rates of cholera. The first British epidemic resulted in 32,000 deaths. The second epidemic started in 1848. At that time there were no effective treatments for cholera. Numerous unusual medications and physical treatments

(Autotype from a Presentation Portrait, 1856, and Autograph facsimile.—B. W. R.)

Figure 4.1 John Snow

were attempted such as bleeding, hot poultices, enemas of every description and potions containing chemicals including ammonia, pepper, turpentine, creosote, bismuth, sulphuric acid, opium and mercury.

Legacy

John Snow published his observations on cholera in a pamphlet of 1849 (Snow, 1849, p. 8) and this was updated and expanded, using British and foreign research, in his 1855 book (Snow, 1855).These reports were self-published but in 1857 his findings were published in a reputable medical journal (Snow, 1857) This discussion of his work is based on the writings of medical historians (Hempel, 2007; Jackson, 2014; Bynum, 2008; Porter, 1997) and epidemiologists (Last, 1983; Donaldson and Donaldson, 1983; McMahon, 1970; Paul, 1973).

On completing his apprenticeship Snow walked from Newcastle to London in 1836 to pursue his training. His humble background and lack of finance made entry to the profession through university impossible. But after apprenticeship he was able to study at the Hunterian School of Medicine (mainly anatomy) at Windmill Street and obtain experience at the Westminster Hospital. He lived frugally and studied hard.

In May 1838 he passed the Membership of the Royal College of Surgeons (MRCS) and in October 1838 he was admitted as a Licentiate of the Society of Apothecaries (LSA). In 1839 he passed the Bachelor of Medicine (MB) examination of the University of London and in 1844 his thesis was accepted for Doctor of Medicine (MD). These qualifications enabled him to do independent medical practice and obtain an income. He became a Member of the Royal College of Physicians in 1850 (MRCP).

A main source of work, which he almost stumbled upon by accident when he saw that an unscientific non-medical practitioner was in demand, was as an anaesthetist using ether and new techniques which had come from America. He used his physiological and chemical knowledge to develop this skill, later using chloroform after James Simpson's reports and was approached to give anaesthesia to Queen Victoria to ease the delivery of her eighth and ninth child, which he did successfully. It has been estimated that his anaesthetic practice only yielded an annual income in modern terms of around £60,000. He was not a fashionable wealthy London physician, although he had royal connections. He gave his medical services to many poor patients and institutions without payment.

His interest in advancing medical knowledge was apparent with his membership of the Westminster Medical Society which met weekly and listened to members' presentations. In the 1848 cholera epidemic he treated patients and tried to identify the cause of the condition. In the terrible conditions in which he saw the poor lived with overcrowding, poor nutrition, poor sanitation and polluted water, he sought to identify an identifiable cause for the disease. In some areas, such as Albion Terrace, Battersea, he saw very high rates of cholera while similar nearby locations were much less affected. The water supply seemed to be a factor but the evidence was not clear cut. Identifying the water sources for families and individuals in overcrowded areas was not easy. Edwin Chadwick and others had tried to improve the living conditions of the working classes and by 1848 it was claimed that Joseph Bazalgette's sewer construction programme had abolished 30,000 cesspools in London but directed their content into the Thames, which often resulted in an unpleasant stench in the capital. When the second epidemic died down in 1849, the national death toll was 52,000.

Snow was a founder member of the London Epidemiological Society in 1850. He and likeminded doctors and statisticians saw the value of using good mortality and morbidity data to study disease. They did not have to wait long as the third cholera epidemic came from the Baltic via Newcastle and to London in 1854. The opportunity for a large natural experiment to test whether polluted drinking water was the culprit arose as two different water companies supplied the 300,000 population of south London. The Southwark and Vauxhall company obtained water from the sewage-contaminated lower Thames while the Lambeth Water Company obtained its supply from the upper unpolluted reaches of the Thames at Thames Ditton. Using the detailed information of weekly cases and deaths from local authorities at ward level, Snow began to investigate where cases occurred. The results showed that 315 cases per 10,000 houses occurred in houses supplied by polluted water compared to 37 per 10,000 houses supplied by water from the upper Thames.

Further evidence also came from a smaller detailed study of cases around the Broad Street pump in Soho in August/September 1854 when over 90 cases occurred within a 10-day period. This was mapped by Snow and showed clearly that cases were almost exclusively in households supplied by the Southwark and Vauxhall Company (see Figure 4.2). In effect, people getting water from the polluted part of the Thames had 14 times more fatal attacks of cholera than those getting their supply from the purer source (Last, 1983). It was also important to observe that some populations located near the pump, such as employees of a local brewery work and inmates of a small prison, were not affected. They obtained their potable water from deep artesian wells on their premises. A few cases also were identified who had moved to live some distance away but preferred to get water collected from the pump. These findings were published (Snow, 1855) but not universally accepted. The famous story of Snow attending a local meeting which resulted in the removal of the handle from the Broad Street pump is well

Figure 4.2 Map of cholera cases in the area of the Broad Street pump

known, and while it had a beneficial effect, it also appears that the epidemic was beginning to decline at that time for other reasons. However, over the next decade the medical and political establishment gradually accepted the evidence that Asiatic cholera was a water-borne gastro-intestinal disease. Surprisingly, many important figures including Florence Nightingale and Edwin Chadwick remained committed to the miasmatic theory of spread. Leading doctors such as John Simon and William Farr later accepted the theory and to some extent claimed credit for proving it with limited acknowledgement of Snow's work.

The fourth British epidemic came in 1866 and showed that there was still much to be done to provide clean water to British houses. It was not until 1883 that the German microbiologist Robert Koch identified micro-scopically the bacterium *Vibrio Cholera* – the organism responsible for the disease. Later research showed that it multiplies in the gut and causes gut cells to pump out fluid laden with mucus – causing 'rice water stools' and dehydration. Rehydration and electrolyte replacement are the mainstays of treatment.

Last (1983, p. 98) has stated that Snow's 1855 publication (Snow, 1855) can be regarded as the first definitive working text on epidemiology and also contained an explicit statement of the germ theory 30 years before Koch discovered the vibrio.

Implications for contemporary policy

Mapping of disease outbreaks is important and involves detailed observations and painstaking personal enquiries. It has been said that research is 99% perspiration and 1% inspiration and this seems true in Snow's work. During his lifetime others propounded his ideas without acknowledging his work and even his obituary in *The Lancet* was brief and ignored his work on cholera. Expecting fame and fortune for doing effective health work is perhaps unrealistic. Snow never obtained Fellowship of the Royal College of Physicians or the Royal College of Surgeons (and is therefore not mentioned in Munk's Roll or Plarr's Lives of the Fellows). However, in 1854 he was made the annual orator of the London Medical Society, the successor to the more local Westminster Medical Society. Like Edward Jenner it was often those abroad who recognised the value of his work. In 1849 the French government awarded him a prize of 150,000 francs for his 1849 paper and in Germany his book was translated and widely circulated.

In later years his health was poor and he modified his abstemious lifestyle by taking some meat and a small amount of wine. A lifelong bachelor he passed away following a stroke at the early age of 45 and is buried at the Brompton Cemetery. At the time of his death he was tackling another public health problem, the adulteration of bread with alum and possible links with rickets in children.

Snow's work has been recognised in America and his fairly simple approach, which yielded vital truths, was used to support the claim that epidemiology should be more generally recognised as one of the basic sciences of medicine (McMahon, 1970).

It is also important to appreciate that by identifying the cholera organism and preventing disease, many potential patients were spared unpleasant but ineffective treatments and sufferers were given effective supportive care.

In 2013 *The Lancet* took the unusual step of making a correction to Snow's brief 1858 obituary to emphasise that he made major contributions to epidemiology and had done visionary work in deucing the mode of transmission of epidemic cholera.

Conclusions

John Snow came from humble beginnings, worked hard for over 10 years to qualify as a doctor who did not pursue his profession to gain personal wealth but made major contributions to public health and anaesthetics with the pioneering use of ether and later chloroform to reduce the pain of surgery and childbirth.

He has been described as the father of modern epidemiology because of his investigations linking the spread of cholera to polluted drinking water. His personal studies were in the London area but he used information from this country and abroad to support his theory that water rather than airborne miasma ('bad air') was the method by which cholera was spread; as such he was a major contributor to modern public health.

Bibliography

Bynum, W. (2008) *The History of Medicine*. Oxford: OUP.

Donaldson, R. J., and Donaldson, L. J. (1983) *Essential Community Medicine*. Lancaster: MTP.

Hempel, S. (2007) *The Strange Case of the Broad Street Pump*. Berkeley: University of California.

Jackson, M. (2014) *The History of Medicine*. London: Oneworld.

Last, J. (ed.) (1983) *A Dictionary of Epidemiology*. Oxford: OUP.

McMahon, Pugh T. F. (1970) *Epidemiology – Principles and Methods*. Boston: Little, Brown and Company.

Paul, J. R. (1973) An Account of the American Epidemiological Society. *Yale Journal of Biology & Medicine*, 46(1).

Porter, R. (1997) *The Greatest Benefit to Mankind*. London: Harper Collins.

Snow, J. (1849) *On the Mode of Communication of Cholera*. London: Churchill.

Snow, J. (1855) *On the Mode of Communication of Cholera* (2nd ed.). London: Churchill.

Snow, J. (1857, October 17) Cholera and the Water of the South Districts of London. *British Medical Journal*, 864–865.

Useful website

www.johnsnowsociety.org/

5   Sir Joseph Bazalgette

A man of persistence and vision

Alan Page

Introduction

> The majority of the inhabitants of cities and towns are frequently uncon-
> scious of the magnitude, intricacy, and extent of the underground works,
> which have been designed and constructed at great cost, and are necessary
> for the maintenance of their health and comfort.
>
> — Bazalgette (1865, p. 3)

Bazalgette's words are as true today as when written. The vast majority
of people do not think about the disposal of daily waste, nor to the conse-
quences of what would result should our sewerage system fail. Unplanned
and un-sequestered waste remains an issue in many parts of the world, so it
is of import to recognise the work of Sir Joseph Bazalgette (Cook, 2005), a
man before his time, who designed a system for London that has lasted well
over 150 years.

The history of sewerage is best expressed by following the experience
of London in which the supply of water to homes and businesses such as
tanneries, slaughterhouses and brewers and the subsequent disposal of their
wastes were one in the same medium. Pope, Swift (Cockayne, 2007) and
Dryden refer to the River Fleet in their poetry and essays:

> Sweeping from Butchers Stalls, Dung, Guts, and Blood,
> Drown'd Puppies, stinking Sprats, all drench'd in Mud,
> Dead Cats and Turnip-Tops come tumbling down the Flood.
>
> Swift (1710) in William Hazlitt (1824)

Bazalgette's legacy and contribution to contemporary public health

It is important to understand the position and effectiveness of London's
sewerage system prior to 1860. It may come as some surprise to know

that prior to 1815 it was an offence to discharge sewage from houses or dwellings direct to the sewer (Bazalgette, 1865; Halliday, 1999). At that time it was considered better to utilise cesspools to store waste, for subsequent later disposal by night soil-men, who collected the waste for storage and later sale as manure. Alongside this London still used communal public toilets, with some such as those on London Bridge being built so as to discharge directly over and into the Thames (Stanwell-Smith, 2010). The problem was that as London's population increased to circa 2.5 million (Krugman and Wells, 2013), so did the number of cesspools and with increasing household appliances so did the necessity for the construction of overflow drains to the sewers, which at first was a permissive right and then controlled by enactment by the Public Health Act 1848 and the Nuisances Removal and Diseases Prevention Act 1848. Within the six years post-enactment some 30,000 cesspits would be systematically removed (Cook, 2001) and all the sewage diverted to sewers that, at that time, either drained directly into the Thames, within the city boundaries or had no outfall and as such the sewage was left to stagnate below ground (Hamlin, 1992). In parallel it is worthy of note that at this time the vast majority of sewers were open ditches. Where underground drainage did exist it was not constructed to a uniform standard (Hamlin, 1992) so that the

> sizes, shapes, and levels of the sewers at the boundaries of the {8} different districts were often very variable. Larger sewers were made to discharge into smaller ones sewers with upright sides and circular crown and inverts were connected to egg-shaped sewers; and egg-shaped sewers with narrow part uppermost were connected to similar sewers having the smaller part downwards.
>
> (Bazalgette, 1865, p. 6)

The severity of the issue, in London and wider country, is summed up by the cholera outbreaks of 1831 and the subsequent 6,536 deaths (Westminster City Archives, n.d.) and 30,000 deaths in England the following year (Smith, 2002; Roberts, 2007); of 1848 (Donaldson, 2008) in which 14,137 died in London (Westminster City Archives, n.d.) and 52,000 people in England (Roberts, 2007) and 1854 in which 10,738 die in London (Westminster City Archives, n.d.); that the level of infant mortality, during the 1830s, in Britain's towns was close to 50% with the leading causes diarrhoea, dysentery, typhoid and cholera (Hart-Davies in Halliday, 1999); and by the "Great Stink" of 1858 (Halliday, 1999, *The Times*, 1858) which highlighted the effect on Parliament in an article of 18 June:

> the intense heat had driven our legislators from those portions of their buildings which overlook the river. A few members, bent upon

investigating the matter to its very depth, ventured into the library but they were instantaneously driven to retreat, each man with a handkerchief to his nose.

The Hansard record of the time (Hansard, 1858) tables a question from Mr R. D. Mangles (MP) on the proposed action to improve the state of the Thames. He points to the considerable delay in finding solutions and the "completely futile" actions of the Metropolitan Board of Works, which had held six separate Metropolitan commissions of sewers between 1847 and 1855 (Bazalgette, 1865). This period included the "pipe-and-brick" sewers war of 1852–1854 (Hamlin, 1992), in which partisan positions on the most effective form of sewerage mechanism stymied any real progress on the development of a cohesive system for London.

Despite the problems that the Thames and London faced, Bazalgette faced considerable opposition from those that believed that it was bad air that caused disease (Halliday, 2001) and thus intervention in water and sewage would not solve the problem. John Snow, at the time, noted that the debate surrounding the questions of causation were often acrimonious, stemming in the main from positions on political or economic ordering of society (Smith, 2002) and with the view that the poor were feckless and bought to themselves the very ills that they faced (Smith, 2002). With this in mind Bazalgette also faced opposition from the financiers and liberals, who favoured a free market (Smith, 2002), and thus felt that it was not the government's place to build a public system. The *Economist* editorial of 13 May 1848 declared that

> suffering and evil are nature's admonitions; they cannot be got rid of and the impatient attempts of benevolence to banish them from the world by legislation . . . have always been productive of more evil than good.

Added to this was the ongoing conflict between the reformers led by Chadwick, who believed that civil engineers, such as Bazalgette, created "hyper-expensive" solutions in collaboration with "irrational" quangos that represented the format of Local Government at the time (Hamlin, 1992). Thus Bazalgette faced opposition from the liberal elite, public health reformers who took a differing position; those supporting miasma theory and even those that opposed the type of sewerage pipework and system advocated (Hamlin, 1992).

But for the "Great Stink" and the return of cholera in 1848 and 1854, which some in the political elite recognised as a threat to the wealthy (Budd, 1849; Smith, 2002), one does wonder whether a metropolitan sewerage system would have come to pass particularly when the costs were estimated at over £3.5 million (Roberts and Arnold, 2007) which at today's value would be an estimated cost of £400 million. Yet by 1891 *The Times* Obituary column (16 March 1891: 4) highlights Bazalgette as

> That great, far-sighted engineer, who probably did more good, and saved more lives, than any single Victorian public official. . . . Of the great sewer that runs beneath, Londoners know, as a rule, nothing, though the Registrar-General could tell them that its existence has added 20 years to their chance of life.

Implications for contemporary policy or practice

Planning a sewerage system for a metropolis that already exists is inevitably fraught, particularly where discharge is to a tidal river. Questions arise as to the amount of sewage created and is the flow uniform across time; do you have a separate sewage and storm water system; what are the minimal invert levels required for the sewers and where can the sewage outfall occur without tidal backwash returning the sewage to the city? There was no national or international comparator, no history or example to rely upon, all of these factors needed to be tested in situ and over time to establish the most effective method of sewage outfall. That this was done at all is a testament to Bazalgette's determination but that it has lasted to date despite considerable increases in population and sewage flow highlights his vision.

As London's Chief Municipal Engineer to the Metropolitan Board of Works, Bazalgette as noted by Boulnois was both meticulous and had a sharp eye for detail (Halliday, 1999). In starting the works in 1859 and completed in 1875 (Westminster City Archives, n.d.), he specified Portland cement in the construction of the sewerage system because of its higher strength, the first time the product had been used in any large-scale system; but more important he introduced a stringent quality control mechanism to overcome the dangers of poor batches (Halliday, 1999). As both designer, with Colonel William Haywood, and project manager he designed and implemented a system through a complex network of sewers, many of which still serve London today, and four huge pumps to discharge London's waste to a point beyond which the tide would return the sewage to the city. The system that Bazalgette constructed consisted of 1,300 miles of sewers consuming 318 million bricks (Cook, 2005). Again in parallel Bazalgette designed and managed the construction of the Northern Thames Embankment, a place at the time rife with mosquitoes that transmitted malaria (Cook, 2001, 2005) at a cost of £2.5 million. The strategic importance and vision of Bazalgette in this construction project was to enable the northern level sewer, accommodate the underground circle link and improvements to the foreshore (Cook, 2001).

Conclusions

Bazalgette's influence was not limited to London alone nor to drainage and sewerage systems. In his wider consultancy role he advised on the construction of the drainage systems for Hampton Court, Cambridge, Norwich, Budapest and Mauritius and also designed Putney, Hammersmith and Battersea Bridges. The bridges and much of the sewerage systems built in the 18th century are still used as we are well into the twenty-first.

Bibliography

Bazalgette, J. W. (1865) On the main drainage of London, and the interception of the sewage from the River Thames: Excerpt minutes of proceedings of the Institution of Civil Engineers Vol. xxiv. Session 1864–65, pp. 280.

Budd, W. (1849) *Malignant Cholera: Its Mode of Propagation and Its Prevention.* London: John Churchill.

Cockayne, E. (2007) *Hubbub: Filth, Noise and Stench in England.* New Haven: Yale University Press.

Cook, G. C. (2001) Construction of London's Victorian Sewers: The Vital Role of Joseph Bazalgette. *History of Medicine,* 77(914), 802–804.

Cook, G. C. (2005) What the Third World Can Learn From Health Improvements in Victorian Britain? *British Medical Journal,* 81, 763–764.

Donaldson, L. (2008) The UK Public Health System: Change and Constancy. *Public Health,* 122(10), 1032–1034.

The Economist editorial 13th May 1848 in Marvin Perry, Myrna Chase, James Jacob, Margaret Jacob, Jonathan Daly and Theodore Von Laue (2016). *Western Civilization: Ideas, Politics and Society Since 1400* (11th ed.). Boston: Cengage Learning.

Halliday, S. (1999) *The Great Stink of London: Sir Joseph Bazalgette and the Cleansing of the Victorian Metropolis.* Stroud: History Press.

Halliday, S. (2001) Death and Miasma in Victorian London: An Obstinate Belief. *British Medical Journal,* 323, 1469–1471.

Hamlin, C. (1992) Edwin Chadwick and the Engineers, 1842–1854: Systems and Antisystems in the Pipe-and-Brick Sewers War. *Technology and Culture,* 33(4), 680–709.

Hansard 18th June 1858 Commons sitting Vol 151, cc 27–40.

Hart-Davis, A. (1999) Foreword to Stephen Halliday. In: *The Great Stink of London: Sir Joseph Bazalgette and the Cleansing of the Victorian Metropolis.* Stroud: History Press.

Krugman, P., and Wells, R. (2013) *Economics* (3rd ed.). New York: Worth Publishers.

Roberts, I., and Arnold, E. (2007) Policy at the Crossroads: Climate Change and Injury Control. *Injury Control*, 13, 222–223.

Roberts, W. C. (2007) Facts and Ideas From Anywhere. *Proceedings (Baylor Medical Center)*, 20(3), 321–333.

Stanwell-Smith, R. (2010) Public Toilets Down the Drain? Why Privies Are a Public Health Concern. *Public Health*, 124(11), 613–116.

Smith, G. D. (2002) Commentary: Behind the Broad Street Pump: Aetiology, Epidemiology and Prevention of Cholera in Mid-19th Century Britain. *International Journal of Epidemiology*, 31, 920–932.

Swift, J. (1710) A description of a city shower: In imitation of Virgil's georgics in William Hazlitt (1824). In: *Select British Poets or New Elegant Extracts From Chaucer to the Present Time*. London: WMC Hall.

The Times. (1891) Obituary: Joseph Bazalgette. 16 March 1891, 4.

Westminster City Archives (n.d.) *Cholera and the Thames*. www.choleraandthe thames.co.uk/ accessed 30 November 2016.

6 George Smith of Coalville ('the Children's Friend')

Campaigner for factory and canal boats legislation

Susan Lammin

Introduction

This chapter looks at the life and achievements of George Smith, also known as 'the Children's Friend'. George was put to work in the tile-works at an early age. He experienced the working life of a child in a harsh industry and his efforts as an adult to make changes were rooted in those early experiences. He had a desire to right the wrongs he suffered and ensure that the commercial system was controlled by statute and that the needs of children were formally recognised and protected.

George Smith was born on16 February 1831 (gravestone Crick grave yard; Crick History Society [CHS], 2009; Bristow, 1999; Green, 2015). He was born in the village of Clayhills, northwest of Tunstall in Staffordshire, into an industrial setting (Hodder, 1896). At the time 'The Potteries' was a densely populated area concerned with the production of china and earthenware (White, 1834). George's father, William, was a brick and tile maker by trade. Both his parents were members of Tunstall's Wesleyan Chapel and George learnt to read and write, from age 4 years, at the Chapel's Sunday School and at Betty Westwood's Dame School (Green, 2015). Dame Schools were run by ladies who for small fees taught basic reading, writing and arithmetic to poorer children (Encyclopedia Brittanica); Betty was a Primitive Methodist (Green, 2015).

At the age of 7 he was put to work in Peake's Tileries (Enyon, 2014). Peake's works were the largest of the five works in the area. By the time George started working they were using steam-driven engines for crushing and preparing the clay (Ward, 1843; Birks, 2008). His job was to carry the clay and empty the kiln. Child labour in the brick and tile yards was common practice (Scrivens, 1841). The regime was harsh for both young boys and girls (Green, 2015). George Smith himself told how at the age of 9 he had to carry 40 lb of clay at a time from the pug mill to the moulders, around 5 tons of clay in a day. On other occasions he carried bricks from the moulding room to the kilns. For a day's work of 13 hours he received sixpence (Smith, 1869; Smith in Miller, 1891; Green, 2015). In 1841, Scrivens had found children working in the industry unschooled and he believed the primary cause was

> sending children at too early a period of life to labour from morning till night, in hundreds of cases for 15 or 16 hours consecutively, with the intermission of only a few minutes to eat their humble food . . . and where they acquire little else than vice, for the wages of 1s. or 2s. per week, whereby they are necessarily deprived of every opportunity of attending a day or evening school.
>
> (Scrivens, 1841, Education and Schools, para. 30)

George Smith was unusual in seeking an education and he attended night school. When he was promoted to watch the kilns two nights a week, he used the extra shilling a week to buy books and pay for an evening at the night-school (Hodder, 1896; CHS, 2009; Enyon, 2014); the Wesleyan Chapel in Tunstall ran a night-school from 1834 to 1874 (Jenkins, 1963). He also kept up his membership of the Primitive Methodists and became a Sunday School teacher himself. At 23 years old he left 'Peake's Tileries' and worked at another brickyard in Ladderedge. While working there he discovered clay at Reapsmoor, and he set up his own works making blue engineering bricks, glazed ware and roof tiles (Baggs et al., 1996; Browne, 2014).

It was at Reapsmoor that he first had the freedom to introduce working-practices according to his conscience. He was open about his motivations, which were based firmly on his religious beliefs and the urge to improve the conditions of women and children in the industry. Therefore, he would not employ women and girls, or boys under 13, and would not let boys work overtime or on Sundays. About 1856 he started a Methodist Sunday School locally (Green, 2015). His approach created a degree of resentment in the area as it meant poor families were losing incomes from the women, girls and young boys.

Unfortunately, his yard was too isolated, the cost of carrying coal to the kilns was too high and the business never really flourished; so he took a job managing a works at Humberstone, Leicestershire (CHS, 2009). While in Leicestershire he discovered a works in the northwest of the county that was only making drainpipes and bricks for a mine. George identified the potential to make more valuable tiles and decorated bricks. He was to have rented the works but once the Whitwick Colliery Company realised how good the clay was they reneged and retained the business themselves and offered him the job of manager at £75 per year; so now, in 1859, George moved to Coalville (Baker, 1983b; Enyon, 2014). This move allowed him to adopt the soubriquet of 'George Smith of Coalville'.

George ran the Whitwick Colliery Company's Terra Metallic Tileries, Ornamental White Brick and Pipeworks without employing any children under the age of 12, nor any girls or women at all. He would not let the boys he employed do overtime or work on Sundays. Yet the business was successful. He put his principles into practice. At the same time he was actively teaching at Sunday Schools in the local chapel and encouraging sponsorship of new Chapel-schools (Green, 2015). In the mid-1800s, in the Potteries, the Sunday schools were the only means by which most children could gain an education (Jenkins, 1963).

In 1864 and 1867 the Government had extended the Factory Acts. The Factory Acts (Extension) Act 1864 incorporated previous factory acts and made the scope of controls wider than just coal mines and textile manufacture, it now included the pottery industry among others. The 1867 Act brought all factories employing more than 50 people under the terms of all existing Factory Acts; forbade the employment of children, young people and women on Sundays; and amended some regulations of previous acts. These laws were to stop people employing children under the age of 8. Children age 8 to 13 could only be employed part-time. They had to spend ten hours a week in school. What is crucial is that the legislation did not apply to workshops employing fewer than 50 people; so did not apply to many brickworks, they were employing children as before. Knowing this, in 1868 George Smith started his campaign against the use of child-labour in brickyards.

George's concerns have been borne out by census data: the employment rates of 10- to 14-year-olds declined only slightly between 1861 and 1871 and between 1871 and 1881 (when provisions were enforced by Factory Inspectors) the employment rate fell by 7% (Nardinelli, 1980, p. 754). The statistics suggest that many children were employed within small businesses (< 50 employees) outside the scope of the 1867 Act.

Once he had achieved changes to the law for brickyards he then moved on to campaign for legislative change to protect children working and living on the canals.

The legacy

George made seven proposals regarding employment of children in the brickyards (1871a, 1871b):

1   To prohibit infant and child-labour in brickyards (and elsewhere);
2   To prohibit the employment of women and girls in brickyards;
3   No one under 12 years to be permitted to work in brickyards, and only then if they can read, write and do arithmetic;

4   Maximum working hours between 8 and 10 hours per day and for the youngest employees (12–14 years) working on alternate days (part-time working);

5   Official supervision of the health and treatment of all minors; with punishments for those maltreating children at work;

6   To include all 'brickworks and tileries' (small businesses with less than 50 employees and large businesses) under the scope of the Factories and Workshop Act; and

7   To have factory inspectors experienced in brick-making and have a universal inspection service so all brickyards were inspected. (Not just 100 out of 2,825 as happened at the time.)

George began to write letters. He wrote to the newspapers. He wrote to the National Association for the Promotion of Social Science. At first it seemed that nobody would listen to him. He wrote to the Factory Inspector, probably prompting official help in his cause. This help came from Robert Baker, one of Her Majesty's Inspectors of Factories. Dr Baker was a qualified doctor who had become one of only two Factory Inspectors in 1858 (Lee, 1964). His role was to enforce legislation and recommend changes to the law; these recommendations appeared in the six-monthly reports of the Inspectors of Factories or in evidence to Parliamentary Commissions (Lee, 1964). Robert Baker visited the Coalville works and afterwards fully and sympathetically represented George's views in his next official Report:

> In one case a boy of 11 yrs was carrying 14lbs weight of clay upon his head, and as much more within his arms, from the temperer to the brickmaker, walking 8 miles a day upon the average of 6 days.
>
> (The Graphic, 27th May 1871)

George Smith was instrumental in having the Factories Act (Brick and Tile Yards) Extension Bill [which became the Factory Acts Extension Act (1864) Amendment Bill of 1871] debated in Parliament, clause 5 of which read,

> After the first day of January, one thousand eight hundred and seventy two, no female under the age of sixteen years and no child under the age of ten years, shall be employed in the manufacture of bricks and tiles.
>
> (Smith, 1871a, p. 112)

He also claimed a champion in J A Mundella MP. Mundella spoke in the House of Commons in favour of change and was of the view that "ignorance, vice and immorality prevail to a greater extent amongst employees in brickyards than in any other trades"; would be likely to persist unless something was done by the Government: "the time has come when children employed in our brickyards should have extended towards them a helping hand" (Smith, 1869). During one parliamentary debate in 1875 Mr Mudella stated: "in one district no less than seven Acts of Parliament were in force . . . ; but the brick-yard . . . children were under no inspection whatever" (Hansard, 1875).

George Smith's early efforts brought with it some enmity from proprietors of brickyards. His unpopularity was not limited to businessmen; some working families around Coalville resented losing income from the women and young children of the family (Hodder, 1896). This problem arose because Coalville had been established to serve the colliery and subsidiary trades; there was little alternative employment initially. Women eventually found employment in other industries, for example in manufacturing; in 1872 T. & J. Jones set up a factory manufacturing elastic web, and five years later Walter Brown set up Boot and Shoe factory in the town (Baker, 1983a; Royle, 1978).

Regardless of some antipathy, he also took up the cause of the barge boys and girls. George had been close to the canals during his working life; Peake's works were only 200 yards from the Trent and Mersey Canal (Girdlestone, 1845) and the yards relied on the canals to transport coal and finished product. His appeal to improve conditions for canal children was effectively launched at the 1874 Sunday School Anniversary of Moira Primitive Methodist chapel where he used the occasion to describe the terrible conditions in which they lived and the lack of access to education (Green, 2015). His employers were unimpressed with his new campaign (Royle, 1983 p. 20). George left his post as works manager and had to endure reduced circumstances.

Implications for contemporary policy and practice

As George Smith successfully influenced policy, the means by which he achieved that success is worthy of note. He was a working-man with limited education who succeeded by his own efforts. His approach was a mixture of letter writing (Smith, 1871a, 1871b, 1871c,1873a, 1873b), lobbying and publishing emotive books (1875a, 1881, 1882) and pamphlets (1875b) to describe what was happening. He wrote (1871a, p. 81):

> Kicks, cuffs, over-hastening, and oaths and curses . . . are the too fre-quent modes of impelling to work. The . . . shivering, cowering, scared looks of many of the children, are things not to be imagined. I . . . have seen, . . . the black eye, the unhealed sore, the swollen head, the bruised body, in . . . very little children, that proclaimed sorrowfully their expe-rience . . . cruelty, murderous violence . . . , and punishment within, not an inch, but a hair's breadth of life.

He had no powerful associates or mentors and was not politically active through party-politics so he had no established network or affiliations when he began to campaign. However, he did attract an influential advocate in HM Factories Inspector Robert Baker. The contribution of Dr Baker's recommendations in focusing the legislature's mind on the issues is not clear but if nothing else it must have lent credibility to George Smith's published descriptions. The MP, John Mundella, also seems to have read George Smith's book (1871a) and used it as a descriptive basis for his contributions to parliamentary debate.

His methods for seeking change on the canal boats were similar. He published books and pamphlets and wrote to the newspapers to raise awareness:

> I have now very respectfully to claim the attention of the public, through the press, to . . . the boatmen and their children, At a rough estimate there are 100.000 of canal boatmen in active employment in England alone . . . , with their families . . . well on half a million. It is no trivial or small interest . . . that these boatmen, their wives, and families . . . pass their lives . . . on their boats . . . [the] cabin . . . is a miserable hole, averaging 6 feet by 7 feet 6, by 4 feet 6, . . . and in such places will be found from four to seven persons, over and above fire-stove, seats, beds, pots, and other necessaries. . . . the children of these 100,000 boatmen not more . . . than 2000 will . . . attend either day school or Sunday school. Of the boatmen . . . 2 per cent as able to read and write . . . it seems monstrous that with the Factory and Workshop Act of 1871, passed and in activity, it should be possible to have women and children herding together . . . in these boats, so that the Education Act should be inoperable in relation to these boatmen's children.
>
> (Smith, 1873a, Letter to the Editor)

George Smith's main tactic was to write descriptively and publish as often as possible; his 1875 book (*A Cry from the Boat Cabins*) is an overt appeal for legislation and contains numerous reprints of newspaper articles. His books were designed to be sentimental and to raise the indignation of their readers. The books prompted support for the Canal Boat Act of 1878 and the Amendment Act of 1884 (Foster, 2006).

Edwin Hodder (1896, p. 5) described his methods to effect political change:

> For over a quarter of a century his name was daily before the public. When Parliament was in session he was to be found in the forepart of the day in Paternoster Row and Fleet Street among publishers and press men; and in the evening he was almost as regularly in the lobby as the

Speaker was in the chair, of the House of Commons . . . he was known, respected and on good terms with the highest in the land.

That George Smith managed to keep the issues in the press is evident; as an excerpt from the *Evening Express* shows:

George Smith of Coalville – The Central News says George Smith of Coalville, who has been in failing health for several months, and is still confined to his house, has written a long letter to the Home Secretary urging him to include the Canal children within the scope of his Factory Bill.

(*Evening Express*, 1895)

This article was printed three months before his death on 21 June 1895. Even when he was ill he kept writing to get his views heard.

Conclusion

George Smith was largely responsible for initiating important reforms in conditions in the brickyards and on the canals of 19th-century Britain. The movement for reform until then had focused on factories, especially textile mills (Kydd, 1857). He was responsible for lobbying for the regulation of employment of young persons and women in brick and tile works. His efforts saw tileries and brickyards included within legislation and this had a demonstrable impact on the age at which children were legally employed in the industry. The Factory and Workshop Act 1878 included brickyards and tileries within the scope of the law. Greater protection was afforded to children: children younger than 10 could not be employed anywhere and between the ages of 10 and 14 they could be employed only for half days (and must attend school). The later Factory and Workshop Act 1901 made the minimum working age of children 12 years . . . so George Smith's own target for the minimum working age was achieved six years after his death.

He initiated the passing of the Canal Boats Act of 1877. This Act was the first to regulate living conditions on canal boats, and to introduce compulsory regulation of vessels used as living accommodation. The 1877 Act gave power to registration authorities to inspect boats and to restrict the number of people who could live on board. However, the legislation simply permitted this to happen rather than required it to happen. Again George Smith sought reform and the Act was amended in 1884.

George's efforts went into writing and describing problems as he saw them, proposing remedies and above all persisting in his message until he achieved his goals.

Bibliography

Baggs, A. P., Cleverdon, M. F., Johnston, D. A., and Tringham, N. J. (1996) Alstonefield: Fawfieldhead. In: *A History of the County of Stafford: Volume 7, Leek and the Moorlands*, ed. C. R. J. Currie and M. W. Greenslade (London, 1996), pp. 27–31. British History Online www.british-history.ac.uk/vch/staffs/vol7/pp. 27-31 accessed 7 June 2016.

Baker, D. W. (1983a) *Coalville: The First 75 Years (1833–1908)*. Leicestershire, UK: Leicestershire Libraries and Information Service.

Baker, D. W. (1983b) *The Leicestershire Historian: Coalville 150 Issue*. Vol. 3, No 1.1982/3. www.le.ac.uk/lahs/downloads/LeicestershireHistorian-Vol.3No.1-1982-83.pdf

Birks, S. (2008) *Brickworks of Stoke-on-Trent and District: 'Peakes Tunstall Tileries; Thomas Peake; John Nash Peake.'* www.thepotteries.org/brickworks/tunstall_tileries.htm

Bristow, A. (1999) *George Smith the Children's Friend*. Chester: Imogen.

Browne, G. (2014) *Visionary Did Much to Improve the Lives of Working Children*. www.leek-news.co.uk/Visionary-did-improve-lives-working-children/story-24559623-detail/story.html #ixzz4

Crick History Society [CHS] (2009) *Crick Village: George Smith*. www.crick.org.uk/smith.html

Enyon, T. (2014) *George Smith—the Children's Friend*. www.alderoak.co.uk/26101/37501.html

Evening Express (1895, March 9) p. 3. http://newspapers.library.wales/view/3251909/3251912/65/ John%20Smith

Foster, E. (2006) *The Power of the Word: How Writings About Boatpeople Affected Perceptions*. www.rchs.org.uk/trial/3-2%20Word%20power.pdf

Girdlestone Charles (1845) *Commissioners for Inquiring Into the State of Large Towns and Populous Districts: Letters on the Unhealthy Condition of the Lower Class of Dwellings, Especially in Large Towns; Founded on the First Report of the Health of Towns Commission, With Notices of Other Documents on the Subject, and an Appendix, Containing Plans and Tables From the Report*. London: Longman, Brown, Green, and Longmans, p. 26. https://books.google.co.uk/books?id=NRAzAQAAMAAJ&pg = PA26&lpg =PA 26&dq=Peake%27s+Tileries&source=bl&ots=Ubo72eRmpf&sig=1-NujNzHRVZh0m8R RpmFX yJt C7 U&hl=en&sa=X&ved=0ahUKEwja_o-kmJPNAhVKLsAKHQF-Bbo4ChDoAQg4MAQ#v=onepage&q=Peake%27s%20Tileries&f=false

Graphic, The (1871, May 27) *The Brickyard Children*. www.old-merseytimes.co.uk/georgesmith.html

Green, M. (2015) *George Smith of Coalville*. www.myprimitivemethodists.org.uk/page/george_smith_of_coalville_-_the_childrens_friend?path=0p3p90p

Hansard (1875, February 19) Mundella J. A. Motion for a Select Committee; *House of Commons Debate* Vol 222 Cc556–67. http://hansard.millbanksystems.com/commons/1875/feb/19/motion-for-a-select-committee

Hodder, E. (1896) *George Smith (of Coalville) – 'The Story of an Enthusiast'*. London: J. Nisbet and Co.

Jenkins, J. G. (ed.) (1963) *A History of the County of Stafford: Volume 8*. London, pp. 81–104. Excerpt @ www.thepotteries.org/chapel/005a.htm and full text @ www.british-history.ac.uk/vch/ staffs /vol8/pp. 81–104

Kydd, S. (1857) *The History of the Factory Movement: From the Year 1802, to the Enactment of the Ten Hours' Bill in 1847* (2 Vols.). London: Simpkin, Marshall, and Co.

Lee, W. R. (1964) Robert Baker: The First Doctor in the Factory Department Part II: 1858 Onwards. *British Journal of industrial Medicine*, 21(167). www.ncbi.nlm. nih.gov/pmc/articles/PMC1038352/pdf/brjindmed00191-0001.pdf

Miller, M. H. (ed.) (1891) *Olde Leeke: Historical, Biographical, Anecdotal, and Archæological: Volume I. Leek, Staffs*. Reprinted from the "Leek Times" (Second Series.) L.P (also: Miller, M. H. (ed.) (1900) *'Olde Leeke'*. Churnet Valley Books, 1900. Reprint.)

Nardinelli, C. (1980, December) Child Labour and the Factory Acts. *Journal of Economic History*, 40(4), 739–755. http://intranet.oit.org.pe/WDMS/bib/virtual/ coleccion_tem/trab_infantil/Child_labor_factor_acts.pdf

Royle, S. A. (1978, June) The Development of Coalville, Leicestershire, in the Nineteenth Century. *The East Midland Geographer*, 7(part 1, 49).

Royle, S. A. (1983) Coalville: Mines and Villages in the 1840s'. *The Leicestershire Historian: Coalville 150 Issue*. Vol. 3, No 1.1982/3, pp. 19–20. www.le.ac.uk/ lahs/downloads/LeicestershireHistorian-Vol.3No.1-1982-83.pdf

Scrivens, S. (1841) *Report to Her Majesties Commissioners on the Employment of Children and Young Persons in the District of the Staffordshire Potteries; and on the Actual State, Condition, and Treatment of Such Children and Young Persons*. www.thepotteries.org/history/scriven.htm

Smith, G. (1869) Letter to the editor: Leicester Chronicle, 4th December 1869. www.old-merseytimes.co.uk/georgesmith.html

Smith, G. (1871a) *The Cry of the Children From the Brick-Yards of England: A Statement and Appeal, With Remedy*. London: Simpkin, Marshall & Co. Leicester, J. & T. Spencer.

Smith, G. (1871b) Letter to the editor: Morning Post, 8th September 1871. www. old-merseytimes.co.uk/georgesmith.html

Smith, G. (1871c) Letter to the editor: *Liverpool Mercury*, 1st May 1871. www.old-merseytimes.co.uk/georgesmith.html

Smith, G. (1873a) Letter to the editor: Cheshire Observer, Saturday 11th October 1873: *"Boatmen and Their Children – An Appeal"*. http://static.premiersite. co.uk/23415/docs/6064609_1.pdf

Smith, G. (1873b) Letter to the editor: Morning Post, 18th March 1873. www.old-merseytimes.co.uk/georgesmith.html

Smith, G. (1875a) *Our Canal Population: A Cry From the Boat Cabins, With Remedy*. London: Haughton.

Smith, G. (1875b) *Our Canal Population: The Sad Condition of the Women and Children – With Remedy: An Appeal to My Fellow Country Men and Women*. Bristol Selected Pamphlets.

Smith, G. (1881) *Canal Adventures by Moonlight*. London: Hodder & Stoughton.

Smith G. (1882) *The Conditions of Our Gypsies and Their Children, With Remedies*, paper read at the Social Science Congress, Nottingham, 26th September 1892.

Ward, J. (1843) *The Borough of Stoke-Upon-Trent, in the Commencement of the Reign of Queen Victoria, Comprising Its History, Statistics, Civil Polity, & Traffic; Also, the Manorial History of Newcastle-Under-Lyme, and Incidental Notices of Other Neighbouring Place & Objects*. London: W. Lewis & Son, p. 101. https://books.google.co.uk/books?id=M6QLAAAAYAAJ&pg=PA100&lpg=PA100&dq=Peake%27s+Tileries&source=bl&ots=geHbsDUFpO&sig=jriUOuNI1_yp_ddsy8hfFOvsseE&hl=en&sa=X&ved=0ahUKEwja_o-kmJPNAhVKLsAKHQF-Bbo4ChDoAQg7MAU#v=onepage&q=Peake% 27s%20 Tileries&f=false

White, W. (1834) *History of the Staffordshire Potteries in History, Gazetteer, and Directory of Staffordshire and the City and County of the City of Lichfield*. Sheffield: R. Leader. https://books.google.co.uk/books?id=mw9Mgl6AHr8C&pg=PA589&lpg=PA589&dq=Clayhills+Staffordshire&source=bl&ots=B2QGBACBjz&sig=zOaiplHIh8uSucC85adr0LhtjY&hl=en&sa=X&ved=0ahUKEwjKmPGpjZPNAhWBJMAKHbG0AI44ChDoAQgwMAU#v=onepage&q=Clayhills%20 Staffordshire&f=false

7   Duncan of Liverpool

The first Medical Officer of Health

Stephen Battersby

Introduction

The choice of subject arose for a number of reasons: Liverpool is the author's home city and where he trained as a public health inspector and Dr Duncan was the first Medical Officer of Health in the country, appointed in 1847. There is also a pub in Liverpool named to commemorate this pioneer who tends to be overlooked in any discussion of 19th-century public health pioneers. More important he worked closely with the country's first Inspector of Nuisances, highlighting a need for closer working between Directors of Public Health and Environmental Health Practitioners in the present public health system.

Duncan did not produce great reports (which is perhaps why he is often overlooked) but he faced particular problems such as scarce resources that have resonance today, even if the issues for public health are different. He recognised that where people lived contributed to their ill-health and that the town of Liverpool had health inequalities that needed to be addressed – the ill-health of the Irish immigrants was due to where they lived; they were not unhealthy because of where they came from or who they were.

Too often in the past the environmental health profession rather resented the involvement of medical practitioners in "public health" which perhaps contributed to the change from "public health inspectors" to "environmental health officers". In 1974 the Medical Officers of Health (MOsH) were transferred to the NHS to "general celebration", thereafter they would only be medical advisers to councils (Johnson, 1983). Also Parkinson (2013) suggested that Thomas Fresh, the first Inspector of Nuisances, was undervalued at the time and his work underreported since by comparison with the "celebrated appointment" of Duncan.

Public health is again a local authority function partly as a result of the recognition of the wider social determinants of health and health inequalities (Marmot, 2010; Wilkinson and Pickett, 2009). So it seems appropriate to recognise one of the important pioneers who sought to change those local environmental conditions that impacted on the health of those lower down the social scale. Duncan sought to address health inequalities on the basis of what was known at the time. Today, when resources are tight and health inequity increasing (but with perhaps greater knowledge), there is a need for those committed to addressing such inequity to use their skills and expertise and worry less about professional backgrounds.

This chapter is the result of a desktop study and literature search including of newspapers of the time via British Library 19th-century newspapers (accessed online) and a review of a Duncan pamphlet (Duncan, 1843) obtained with the help of the staff at The Athenaeum Liverpool.

Duncan: a historical perspective

William Henry Duncan was born in Liverpool in 1805. After graduating in medicine in Edinburgh in 1829, he returned to Liverpool where he established himself in general practice. He became a prominent figure in the social and professional life of Liverpool and lectured in medical jurisprudence at Liverpool Medical School. He contributed to a number of the learned societies in Liverpool. For example in 1830 Dr Duncan was a member of the committee of the Liverpool Mechanics' and Apprentices' Library, which had become a public institution in 1824 (*Liverpool Mercury*, 1830). Such libraries existed before public libraries and this one was founded to help those employed in various trades (Darcy, 1976). Duncan also helped establish the Liverpool Medical Society and the Liverpool Medical Institution (Frazer, 1997) and he was secretary of the latter in 1814 and president from 1836–1838 (Morris and Ashton, 1997). He presented papers and pamphlets to the local societies, particularly on health matters, most notably that delivered in March 1843 to the Literary and Philosophical Society of Liverpool on the physical causes of the high mortality rate in Liverpool (Duncan, 1843). His research coincided with the publication of Chadwick's report 'Inquiry into the Sanitary Conditions of the Labouring Population of Great Britain' (Chadwick, 1842). This in turn led to the formation of the Health of Towns Association and Duncan was secretary of the Liverpool Association. An important consequence of the Liverpool Association was the Liverpool Sanatory Act 1846[1] (a private Act promoted by the Council). This was the precursor to the Public Health Act 1848. The development of an effective Victorian public health movement was partly the result of the efforts of the local Health of Towns Associations which promoted the appointment of inspectors of nuisances (Parkinson, 2014) The Liverpool Association had participants from business, the churches and civic leaders and others in the medical profession, and worked with other associations on the national political scene (Morris and Ashton, 1997).

Liverpool, founded by Royal Charter in 1207, had remained a small settlement with most of the trade with Ireland, and Frazer (1997) says that in the mid-16th century there was only a population of 700 and 13 vessels using the port. The first period of marked expansion came with the Restoration. The first wet dock (the first in the world) was built in 1715 (Liverpool Maritime Museum, 2016a) Liverpool expanded as ports on the River Dee (used primarily as naval ports) declined as the result of silting (Morris and Ashton, 1997). In 1774 Liverpool was seen as "one of the healthiest places in the kingdom" (Duncan, 1843 citing Dr Dobson in Enfield's *History of Leverpool*,[2] 2nd edn, p. 38). The Mersey was deep[3] and wide allowing larger ships to enter the port in numbers and docks meant ships could unload in safe water-filled harbours, allowing for speedy turnaround of ships.

The end of the 18th century and early 19th century was a period of spectacular growth in the importance of Liverpool as a port and trading centre. This growth had in part been the result of the slave trade (or the triangle trade), with ships taking manufactured goods to Africa from all over Britain, transporting slaves to the plantations of the Caribbean and America and returning with the produce from those plantations.[4] Liverpool had 17% of all trade through English ports in 1792 (The Athenaeum, 2016). However it is suggested that by the beginning of the 19th century, 40% of the world's trade was passing through the docks at Liverpool (Liverpool Portal, 2016). The town emerged as a global port based around international trade in salt, slaves, raw material and manufactures during the 18th and 19th centuries (Wilks-Heeg, 2003 cited in Sykes et al. (2013)), and began to vie with London in terms of global connections. The total tonnage passing through the port was 1.56 million tons in 1835 and 2.15 million tons by 1840, doubling again by 1860 (BHO, 2016). Eight new docks were built between 1815 and 1835 (Frazer, 1997).

From 1760 to 1801, Liverpool's population grew from 25,000 to 77,000 but by 1837 there were 203,327 inhabitants. By the end of the 18th century, the town supported ship owners, merchants and a burgeoning professional class including lawyers, bankers and insurance companies (associated with shipping) along with doctors, and clerics, but the main population growth was in unskilled labour. The density of the population in 1837 was 83,415 inhabitants to the square mile and with an annual mortality rate of 3.931%, by comparison Manchester had 73,1221 inhabitants to the square mile and a mortality rate of 3.218% (HMSO, 1839).

Unskilled labourers were needed to build new docks for the ever-expanding shipping trade and to load and unload the cargoes; these were casual labour and low paid. Although reference is often made to the large number of Irish immigrants in the population at this time, which was the case, this was not the result of the Great Famine that occurred in Ireland between 1845 and 1852. Rather it is likely that many Irish came to Liverpool to participate in the construction of the infrastructure associated with the rapid development of the port and/or en-route for the United States so as to improve their economic status. Ireland had largely benefitted from price rises associated with the Napoleonic wars but suffered from the drop in export price levels following Waterloo. From 1815 to the start of the Great Irish Famine between 800,000 and 1 million Irish sailed for North America (UCC, 2016).

The first case of cholera in Liverpool was officially recorded on 17 May 1832. From then until the last case on 13 September 1832, there were 4,977 reported instances of the disease (about 3% of the city's population of 165,775), with 1,523 deaths (a mortality of 31%). Most of the affected, and particularly those who died, were poor, often cellar dwellers and frequently Irish immigrants. Outside London, Liverpool's cholera caseload was the highest in England. Yet Duncan showed this was not a case of ill-health being imported, as their health was often good until they arrived in Liverpool. In a footnote in his key pamphlet Duncan refers to the discussion following his original presentation where he pointed out that that although the Irish residents experienced excessive mortality there could little doubt that "the number of Irish leaving Liverpool in bad health much exceeds the number of those who arrive in a similar predicament." He also pointed out that his Irish patients frequently dated their illness from their arrival in Liverpool (Duncan, 1843, p. 61).

Liverpool experienced more "cholera riots" than elsewhere in the UK. Between 29 May and 10 June 1832, eight major street riots occurred, with several other minor disturbances. This anger was directed at the local medical fraternity because of the perception that the victims were removed and killed by doctors to use the bodies for anatomical dissection. The Cholera Riots of 1832 perhaps demonstrate the complex social responses to epidemic disease, as well as the fragile interface between the public and the medical profession (Burrell and Gill, 2005). Yet it seems that Duncan retained the confidence of the otherwise suspicious populace. By September 1832 the number of cases had dropped to such an extent that reports were no longer issued and hospitals closed. The Health Board expressed thanks to Dr Duncan of the Toxteth Park Hospital "for their gratuitous and unceasing exertions on behalf of the sick" (*Lancaster Gazette* etc, 1832). This was before Snow's work on the mode of communication of cholera, published in 1849. However, it seemed that in later outbreaks such as in 1849, Duncan remained uncertain as to how cholera was spread (Frazer, 1997).

Duncan gave evidence to the House of Commons' Committee on the Health of Towns in 1840 on the prevalence of disease among working classes and the high rate of mortality in Liverpool. This evidence was "stigmatised" by some of the public authorities of Liverpool, as was Duncan (Duncan, 1843). In his pamphlet he related the poor health of the working classes to factors associated with their living conditions (Duncan, 1843). He demonstrated the "unhealthfulness" of the Liverpool population by using mortality data for one year whereby the average age of death in the "metropolis was $26^1/_2$, in Manchester it was 19 and Liverpool it was 17." He also estimated that out of a population of 223,054 about 160,000 could be said to belong to the working classes.

Duncan highlighted the problems of Liverpool that led to ill-health, the 6,294 inhabited cellars with 20,168 inhabitants. Many of cellars had no windows. He highlighted the narrowness of streets and that cellar inhabitants were without any place to deposit their "refuse matter". In many instances the inhabitants of the front houses in the "courts" and cellars made use of the conveniences of the courts so that the ashpits "become full to overflowing". While much had been done to improve the provision of drains and sewers, he pointed out that these had been principally main sewers and only very few of the side streets were served. The 25 miles or so of sewers were unevenly distributed, so of 20 miles of streets occupied by the "working population" only about four miles were sewered. He highlighted the problems of overcrowding in the cellars and lodging houses. He found that as many as 30 people were sometimes packed together in the cellars, 30 people where the air supply would barely be adequate for seven.[5] In summary he said that the "vicious construction of the dwellings, the insufficient supply of receptacles for refuse and excrement, the absence of drains and deficient sewerage and the overcrowding" increased the mortality of Liverpool. He showed how the mortality rate was worse than for any other city not least because of the higher cellar population and the greater density of the population. He also highlighted inadequate street cleansing. People died because of inadequate ventilation, they were surrounded by filth and because they were crowded. "Fever cases" were the most numerous among the courts which were worst ventilated, and he showed with his data that the proportion of damp and wet cellars was greatest where the "fever reaches its maximum" (Duncan, 1843).

On Duncan's appointment as MoH the *Liverpool Mercury* having first criticised the Town Council for being selective as to when money was available or not said the appointment of Duncan as medical officer "does credit to the Council". As the newspaper said "that gentleman was one of the first to take up the sanitary question, and has since been one of its most active and persevering advocates" (*Liverpool Mercury*, 1847). Duncan's appointment on a salary of £300 enabled him to continue in private practice. The Home Secretary had then reproved the Committee for fixing such a low salary and after considerable delay Duncan received the appointment from the Council at £750 per year but on the understanding he relinquished private practice and attended to his duties on a full-time basis (*Liverpool Mercury*, 1863).

Duncan reported to the Health Committee and the *Liverpool Mercury* carried detailed reports each time. For example in May 1857 Duncan reported that "owing to a decrease in inflammatory affections of the lungs the deaths in the borough, which in the previous week were 236 fell last week to 192" (*Liverpool Mercury*, 1857). In 1862, as well as Duncan, who by now was suffering ill-health (Frazer, 1997), the borough engineer, building surveyor and inspector of nuisances reported various matters for their departments including identification and remedy of 89 nuisances, examination of 727 houses (containing 2,127 apartments) and that 351 cellars had been put under examination with the view of improving their sanitary condition (*Liverpool Mercury*, 1862).

Duncan died on 23 May 1863 and his death was announced in the *Liverpool Mercury* on 26 May. It said he had been in a delicate state of health for several months though no immediate danger was "apprehended". It continued that Dr Duncan's name had first been brought to notice in connection with the sanitary state of towns by the work he published in 1843. It was not a surprise that on the establishment of the health committee in Liverpool, the "services of the gentleman who had paid so much attention to the health of the town should be solicited in the capacity of medical officer of health".

Between 1847 and 1858 some 80 miles of sewers with 66 miles of "main drains" were built (Frazer, 1997). Duncan and the engineer James Newlands together had resisted pressure from the council for priority to be given for new sewers in the better-class districts, arguing it was a better health policy to sewer and drain as quickly as possible the more densely populated areas. Bye-laws now included provision of space standards for dwellings and window sizes and provisions for the proper arrangements of ashpits and conveniences. All houses had a supply of safe water except the courts that at least had access to a tap (Frazer, 1997 quoting a letter from Duncan to Dr Gairdner of Edinburgh). Baths and washhouses were erected and large numbers were removed from living in cellars. "Habitable" cellars were registered and put under control.

Legacy

It is arguable that without Duncan, Newlands and Fresh and their work in Liverpool, the development of public and environmental health would have been very different.

Liverpool still has problems with housing conditions as the result of major socio-economic changes over the past 50 or more years. There remain areas of older housing, many built to replace the dwellings Duncan encountered. This was despite Liverpool, along with many cities, having had a massive slum clearance programme up until the early 1970s in an attempt to deal with unfit homes. An Audit Commission paper (undated) as part of the Building Better Lives project (Audit Commission, 2009) used Liverpool as one case study. This reports Liverpool City Council as recognising the link between poor quality housing and occupants' health, yet the city has some of the worst health inequalities in the country, and life expectancy for males was the third lowest in the country. To address the problems the Healthy Homes Programme was developed to take a more co-ordinated approach, with all agencies engaged. Initially there was some funding from what was then the Primary Care Trust. So the work Duncan started continues even today.

Implications for contemporary policy and practice

When Edwin Chadwick was preparing his report on 'the Sanitary Conditions of the Labouring Population' (Chadwick, 1842) he inquired of Duncan what resources were available to him. Duncan's response was, "The following is my entire establishment . . . your servant, William Henry Duncan."

Ashton has said (2004) frequently public health practitioners argue for more resources, particularly with current cuts to public expenditure, and "they cannot do what is needed without more dedicated resources" and public health systems can only cope with three or four priorities. Duncan could have been overwhelmed by the cholera, typhus, dysentery and the thousands living in overcrowded cellars, but worked tirelessly to counter the insanitary conditions in Liverpool before his appointment as MoH (as indeed had Fresh) (Parkinson, 2014). Fresh made environmental health interventions before the appointments in 1847 (Parkinson, 2013), but Duncan had also been a campaigner.

Duncan showed public (and environmental) health is about mobilising and working through others. As Ashton (2004) said he mobilised those resources needed to tackle the threats to health. The argument here is that it should not be a matter of one pioneer being somehow superior to another, rather that there is a need to recognise fully the contribution and partnership of those who worked to improve the public health of the "black spot on the Mersey".

What is also important at this time is Duncan demonstrated that while others sought to blame immigrants for the problems facing the City, the root cause of the problem was inequity and the conditions under which the immigrant workforce was living. They arrived healthy but the conditions under which they were living made them ill.

Conclusions

Socio-economic changes have public health implications and public and environmental health professionals cannot operate in a vacuum ignoring such changes.

In 1974 the environmental health function was no longer a statutory appointment. The changes then reflected the notion that the work of EHOs fitted a different model of health from the medical model. The thinking changed again as a different view of public health became accepted, partly as the result of the Committee of Inquiry into the future development of the public health function (HMSO, 1988, p. 1), which said

> in the past the term 'public health' has commonly if mistakenly, been rather narrowly interpreted and associated in particular with sanitary hygiene and epidemic disease control. We prefer our broader definition based on that formulated by the World Health Organization (WHO, 1946).

No single profession can be responsible for public health, and environmental health officers/practitioners are part of the public health movement and workforce. Inequalities in health and the social determinants of health, although not called that by Duncan, remain, and all those interested in addressing these inequalities have to work together with the same commitment as Duncan.

Notes

1 According to Rosen (1993) the first comprehensive sanitary measure passed in England, which gave the town council the power to appoint a medical officer of health, a borough engineer and an inspector of nuisances.

2 How the name of the town was spelt in that publication.

3 Although it has the second-highest tidal range in the UK according to the National Oceanography Centre, which can be 10 m in Spring Tides: www.ntslf. org/about-tides/river-mersey

4 Liverpool became the British Empire's main slaving port. By 1750 Liverpool had overtaken Bristol and London, and the town's ships dominated the trade until abolition in 1807. In the 1790s Liverpool controlled 80% of the British slave trade and over 40% of the European slave trade (Merseyside Maritime Museum, 2016b).

5 Duncan refers to the Inspectors of Prisons who recommended not less than 1,000 cubic feet for every prisoner, that is 28.3 cubic metres; for comparison if we take a room of 6.5 square metres with an average ceiling height, that room would have 15.6 cubic metres (551 cubic feet).

64   *Stephen Battersby*

Bibliography

Ashton, J. (2004) Aphorism of the Month. *Journal of Epidemiology and Community Health*, 58(8), 717.

The Athenaeum website (2016) *History of the Athenaeum*. http://theathenaeum.org. uk/the-athenaeum/history-of-the-athenaeum/ accessed 16 June 2016.

Audit Commission. (2009) *Building Better Lives – Getting the Best From Strategic Housing*. London: Audit Commission. www.yumpu.com/en/document/ view/2409462/building-better-lives-getting-the-best-from-audit-commission accessed 31 May 2016.

Audit Commission (undated) *Liverpool – Building Better Lives Case Study*. London: Audit Commission.

BHO (1911) Liverpool: Trade, population and geographical growth. In: *A History of the County of Lancaster: Volume 4*, ed. William Farrer and J. Brownbill. London, pp. 37–38. British History Online www.british-history.ac.uk/vch/lancs/vol4/ pp. 37–38 accessed 22 June 2016.

Burrell, S., and Gill, G. (2005, October) The Liverpool Cholera Epidemic of 1832 and Anatomical Dissection – Medical Mistrust and Civil Unrest. *Journal of the History of Medicine and Allied Sciences*, 60(4), 478–498, doi:10.1093/jhmas/ jri061

Chadwick, E. (1842) *Inquiry Into the Sanitary Conditions of the Labouring Population of Great Britain and On the Means of Its Improvement*. London. www. deltaomega.org/documents/ChadwickClassic.pdf

Darcy, C. P. (1976) *The Encouragement of the Fine Arts in Lancashire 1760–1860*. Chetham Society: Manchester.

Duncan, W. H. (1843) *On the Physical Causes of the High Rate of Mortality in Liverpool, Literary and Philosophical Society of Liverpool* (Copy obtained from the library of The Athenaeum Liverpool).

Frazer, W. M. (1997) *Duncan of Liverpool*. Preston, Lancashire: Carnegie Publishing.

HMSO. (1988) *Public Health in England – The Report of the Committee of Enquiry Into the Future Development of the Public Health Function, Cm 289*. London: HMSO.

HMSO. (1839) *First Annual Report of the Registrar General of Births, Deaths, and Marriages in England*, Presented to both House of Parliament by Command of Her Majesty, London, England. https://books.google.co.uk/books?id=5vRDA QAAMAAJ&pg=PA81&lpg=PA81&dq=Population+of+Liverpool+1560&so urce=bl&ots=dADErm2vu0&sig=C1FPE3xM8BKmDWO8t6kVsPNiF9k&hl =en&sa=X&redir_esc=y#v=onepage&q=Population%20of%20Liverpool%20 1560&f=false accessed 31 May 2016.

Johnson, R. (1983) *A Century of Progress – History of the Institution of Environmental Health Officers 1883–1983*. London: IEHO.

Lancaster Gazette and General Advertiser, for Lancashire, Westmorland, &c. (1832) "The Cholera" Saturday, September 29, 1832; Issue 1633.

Liverpool Mercury etc (1830) "Sixth Annual Report of the Liverpool Mechanics' and Apprentices' Library" February 19, 1830, Issue 981.

Liverpool Mercury etc (1847) "Liverpool Town Council" January 8, 1847, Issue 1862.

Liverpool Mercury etc (1857) "Health Committee", Friday May 29, 1857, Issue 2988.

Liverpool Mercury etc (1862) "Health Committee", Friday January 10, 1862, Issue 4342.

Liverpool Mercury etc (1863) "Death of Dr Duncan, the Medical Officer", Tuesday May 26, 1863, Issue 4771.

Liverpool Portal (2016) www.liverpoolcityportal.co.uk/history/history_index.html accessed 27 June 2016.

Marmot. (2010) *Fair Society, Healthy Lives, the Marmot Review – Strategic Review of Health Inequalities in England Post-2010.* London: UCL, www.ucl.ac.uk/marmotreview.

Merseyside Maritime Museum (2016a) The Port of Liverpool, Information Sheet 34, www.liverpoolmuseums.org.uk/maritime/archive/sheet/34 accessed 26 June 2016.

Merseyside Maritime Museum (2016b) Liverpool and the Atlantic Slave Trade, Information Sheet 3 www.liverpoolmuseums.org.uk/maritime/archive/sheet/3 accessed 26 June 2016.

Morris, M., and Ashton, J. (1997) *The Pool of Life – a Public Health Ealk in Liverpool.* Liverpool: Department of Public Health.

Parkinson, N. (2013, November) Thomas Fresh (1803–1861) Inspector of Nuisances, Liverpool's First Public Health Officer. *Journal of Medical Biography,* 21(4), 238–249.

Parkinson, N. (2014) The Health of Towns Association and the Genesis of the Environmental Health Practitioner. *Journal of Environmental Health Research.* London: CIEH.

Rosen, G. (1993) *History of Public Health.* Baltimore: John Hopkins University Press.

Sykes, O., Brown, J., Cocks, M., Shaw, D., and Couch, C. (2013, December) A City Profile of Liverpool. *Cities* (International Journal of Urban Policy and Planning) 35, 299–318.

University College Cork (UCC) *Irish Emigration History.* www.ucc.ie/en/emigre/history/#_ftn3 accessed 16 June 2016.

Wilkinson, R., and Pickett, K. (2009) *The Spirit Level-Why Equality Is Better for Everyone.* London: Allen Lane.

World Health Organization. (1946) Preamble to the Constitution of the World Health Organization as adopted by the International Health Conference, New York, 19–22 June, 1946; signed on 22 July 1946 by the representatives of 61 States (Official Records of the World Health Organization, no. 2, p. 100) and entered into force on 7 April 1948. www.who.int/about/definition/en/print.html

8 Margaret McMillan

Advocate and practitioner of improvements in children's health

Susan Lammin

Introduction

Margaret was born in the mid-Victorian era, and her agitation for legislative change was during the early 1900s, when political pressure for social change was building and new political affiliations were being formed in Britain. This chapter looks at the life and achievements of Margaret McMillan in chronological order. In this way the context of economic, social and political changes are more easily followed.

Margaret McMillan was born on 20 July 1860 in the USA. When she was 5 years old her father and a sister died. Her mother then returned to Scotland, bringing Margaret and her older sister (Rachel) with her. Margaret was educated at Inverness High School.

When her mother died, Margaret was 17 years old and had to be prepared for employment. She left home to be 'a governess' (Ewan et al., 2006). Becoming a governess was the only acceptable way of earning money open to a middle-class woman of the time (Hughes, 1993). Her elder sister had been kept at home to help care for her aging grandmother but, after she died, both sisters lived together and worked in London. This early history is documented in Margaret's writings and specifically in the biography of her sister (1927). During much her adult life, Margaret and her older sister Rachel (1859–1917) lived and worked closely together and Rachel seems to have been the catalyst for Margaret's political views and reform effort.

In the year before she came down to London, Rachel had visited a cousin who took her to hear a sermon by Rev John Glasse, a leading Christian Socialist in Edinburgh (William Morris, 1888). The melding of religion and politics was typical of the time (Jones, 1968). Rachel was also introduced to John Gilray and went to several socialist meetings (McMillan, 1927). Gilray gave Rachel copies of *Justice*, the weekly newspaper of the Social Democratic Federation (Cole, 1905) and 'An Appeal to the Young' (Kropotkin, 1880).

Kropotkin's article struck a chord; he had been working towards a social revolution in Russia (Cham, 1989). 'An Appeal to the Young' was aimed at young people preparing to enter the professions. It was first published in *Le Révolté* and then issued as a pamphlet. Kropotkin aimed to awaken a social conscience:

> you ascend into a foul atmosphere by the flickering light of a little ill trimmed lamp; you climb . . . flights of filthy stairs, and in a dark, cold room you find the sick woman, lying on a pallet covered with dirty rags. Pale, livid children, shivering under their scanty garments, gaze at you with their big eyes wide open . . . misery glares upon the family in all its squalid hideousness . . . you will return home . . . saying to yourself, "No, it is unjust; this must not go on any longer. It is not enough to cure diseases; we must prevent them. A little good living and intellectual development. . . . Air, good diet, less crushing toil – that is how we must begin."
>
> (Kropotkin, 1880, p. 4)

Rachel was convinced. When she went back to Inverness she wrote to a friend: "I think that, . . . when these teachings and ideas are better known, people generally will declare themselves Socialists" (McMillan, 1927, pp. 28–29).

Within months Margaret was living with her sister, a woman who was evangelising her newly found socialist ethos. Margaret began attending socialist meetings with Rachel. They both wrote for the *Christian Socialist* magazine. In the evenings they gave free lessons to girls. It may have been these lessons coupled with the political rhetoric that confirmed for Margaret the connection between a worker's physical environment and their intellectual development.

The legacy

Margaret started to develop skills as an orator and she expanded her writing. Her open politicking and speeches were not well received by her employer and Margaret was looking for another position. It was then that Fred W. Jowett of the Social Democratic Federation asked her and her sister to move to Bradford to continue the type of political work they'd been doing (McMillan, 1927).

In Bradford the sisters toured, speaking at Christian Socialist meetings and visiting workers' homes. Bradford's workers suffered falling wages, loss of work and poor living conditions at the time, as the trade in wool-worsted declined (Parkin, 2011). The conditions Margaret witnessed included overcrowded, back-to-back houses built haphazardly and with no drains. Bradford Sanitary Committee's Report (Smith, 1845) had previously documented conditions: for example a single room 15 feet square, occupied by six people: "There are two privies within six feet of the dwellings from whence the excrement overflows. Ashes, refuse and filthy water accumulate with this and send forth an intolerable stench. . . . Various diseases have afflicted (the) parties" (Smith, 1845). Bradford Corporation had started clearing the slums in1877 (Duxbury-Neumann, 2015), under the Artisans' and Labourers' Dwellings Improvement Act 1875, but unfit dwellings still remained for decades.

Margaret was concerned by what she saw: the wretchedness of the slum children and society's apparent obliviousness.

> The condition of the poorer children was worse than anything that was described or painted. It was a thing that this generation is glad to forget. The neglect of infants, the utter neglect almost of toddlers and older children . . . , all combined to make of a once vigorous people a race of undergrown and spoiled adolescents . . . in the 1890s people saw the misery of poor children without perturbation.
>
> (McMillan, 1927, p. 120)

It was to have society recognise the needs of children and to effect changes that Margaret now began to campaign.

Both sisters became more politically active; they joined the Fabian Society, the Labour Church, the Social Democratic Federation and the newly formed Independent Labour Party (ILP) [established 1893]. The ILP's programme of reform included the abolition of child labour. Margaret became focused on improving the physical and intellectual development of children living in the slums.

Margaret and Fred Jowett became the founder members the Bradford Branch of the Independent Labour Party [1893]. Margaret experienced her first personal success at political lobbying when she worked with Dr James Kerr, the medical superintendent of the Bradford School Board, undertaking the first medical inspection of elementary school children in Britain. It was through inspection that they garnered facts. They jointly published a report on their findings and then began to campaign for improvements in conditions. They wanted the School Board to install bathrooms, improve ventilation and supply free school meals. The result was that Bradford became the first School Board in Britain to provide medical inspection and school baths in some of its poorest areas. Dr Kerr organised the school medical service in Bradford before moving to London (BJO, 1941).

Margaret was elected onto the Bradford School Board [1894] (McMillan, 1927 and Bradford archive), as a representative of the Independent

Labour Party. She also wrote books and pamphlets to spread awareness of the issues: *Child Labour and the Half Time System* (1896) and *Early Childhood* (1900). In the census she described herself as an 'authoress' (UK Census, 1901). Unfortunately for Margaret, School Boards were abolished in 1902 and responsibility for education was placed with local authorities, and women could not serve as Councillors.

Margaret returned to London [1902], her sister had already settled in Kent (UK Census, 1901). She joined the Labour Party [established 1900] and was able to work closely with the party leaders. She still sought to improve conditions for poor children and was appointed as the manager of a group of Deptford elementary schools in 1903. She continued to write books conveying her ideas: *Education Through the Imagination* (1904), *Infant Mortality* (1905) and *The Economic Aspects of Child Labour and Education* (1905). Margaret also continued to support Fred Jowett (now a Bradford Councillor) in persuading the local authority in Bradford to provide free school meals. They succeeded in 1904 (Haworth and Haytor, 2015), although in doing so the Council was acting *ultra vires*.

Margaret's party-political efforts and her social-reform work overlapped but the Labour Party did not have a reform programme. The party was willing "to co-operate with any party . . . engaged in promoting legislation in the direct interest of Labour" (Kier Hardy, 1900, p. 31). It was in this environment that the sisters led the campaign for school meals. They were building on Margaret and Fred Jowett's campaign in Bradford. As one of the new Labour MPs, Fred Jowett used his maiden speech to launch the campaign (Gillard, 2003). During the debate, Fred Jowett explained that: " In September, 1904 . . . teachers . . . reported . . . some 3,000 children in the Bradford schools were insufficiently fed" (JHC, 1906). Eventually the new Liberal Government accepted the argument that 'if the state insists on compulsory education it must take responsibility for the proper nourishment of school children' and passed the Education (Provision of Meals) Act 1906. The Act allowed – but did not make it a duty – for local authorities to provide school meals. It made Bradford Council's actions lawful.

The same year, she, Rachel and Katherine Glasier (née Conway) also lobbied Parliament for the compulsory medical inspection and care of school children (McMillan, 1907, 1927, p. 115). Again she was building on her experience from Bradford. The legislation that resulted from their efforts was the Education (Administrative Provisions) Act 1907. Medical inspections did begin (slowly) and the Act empowered Education Authorities to attend to the health and physical condition of elementary schoolchildren.

Lobbying for legislative change wasn't Margaret's only ambition; she was still trying to force the middle and upper classes to recognise the needs of the poor. One outcome of the lobbying for the 1907 Act was a large (£5,000) donation from Mr Joseph Fels, which allowed Margaret and Rachel to open a clinic in Bow [1908]. The clinic was to demonstrate the benefit of medical inspection and care. After a trial-run in Bow, the 'School Treatment Centre' opened in Deptford on 21 June 1910. Margaret reported that immediately, 'the children began to attend in torrents' (McMillan, 1927, p. 120). They expanded into new premises nearby the next year.

Activities in the early days of the clinic demonstrated that ill-health prevention was essential. 'These children come here, are cured and go but in two weeks, sometimes less, they are back again. All these ailments could be prevented; their cause is dirt, lack of light and sun, fresh air and good food' (attributed to Nurse Spiker and quoted in Jarvis, 2013, p. 2). Margaret promoted the clinic in her writing: "forty to sixty children pass in the afternoon to receive treatment [from the nurse]" (MacMillan, 1911, p. 460).

The sisters then opened an overnight camp in the garden of the new clinic. The purpose of the camp was to improve hygiene and so reduce sickness rates. Margaret wrote "Baths are as yet unknown. . . in London schools; we have to go on agitating for them" (McMillan, 1911, p. 463). The McMillans' camp had showering facilities; they made sure that the children were thoroughly washed before going to bed and helped the older girls cook a nutritious breakfast. The camp was a success and was extended to boys, at a different venue, in 1912. Their observations showed that attendance at the camps improved the health and behaviour of the children. The next step was to look at creating a camp-school with lessons in the open-air to remove children from overcrowded classrooms.

The sisters' practical approach to seeking change then extended to younger children. Margaret wrote about her rationale for a nursery "young children . . . [are in] the period in which human life is most susceptible to disease, the price of neglect is paid by society, not only in the high death rate of this period, but in the physically and morally defective specimens who survive" (McMillan, 1919, pp. 153–154). Her political ethos had focused her efforts on the nursery as a remedy for social disadvantage.

They opened the Open-Air Nursery School & Training Centre in Peckham, in 1914. Margaret said "We must plan the right kind of environment for them and give them sunshine, fresh air and good food before they become rickety and diseased" (Stevinson, 1954, p. 8). Within a few weeks of opening there were thirty children at the school ranging in age from 18 months to 7 years. Margaret continued to run the Peckham Nursery and in 1923 became the first president of the newly formed Nursery School Association (NSA).

The clinic and the nursery were both practical efforts to change lives and a chance to gather evidence to demonstrate that those changes were worthy of amendments to national policy. Margaret's nursery was an exemplar heeded by Parliament. Mr George Lansbury asked the President of the Board of Education

> whether he is aware that for many years Miss Margaret McMillan has carried on successful experiments in connection with the provision of nursery schools at Deptford and elsewhere; and whether, in view of . . . the great success which has resulted from them, . . . he will suggest to local education authorities the advisability of establishing similar nursery open-air schools.
>
> (Hansard, 1924)

In a written answer Charles Trevelyan said "I am well aware of the good work which is being carried on at the Deptford nursery schools, and . . . I am prepared to consider sympathetically any proposals" (Hansard, 1924). Margaret actively sought to influence both policy and practice: she served on the London County Council and wrote a series of books that included *The Nursery School* (1919) and *Nursery Schools: A Practical Handbook* (1920).

Just before she died on 29 March 1931, with the financial help of Lloyds of London, she established a new college to train nurses and teachers, the 'Rachel McMillan College' (opened in Deptford on 8 May 1930).

Implications for contemporary policy and practice

The thought that children's development and therefore their adult lives could be improved by adequate food, fresh air, cleanliness and care was a new idea in the 19th century, but the notion of the wider determinants of health are fundamental to environmental health practice now (WHO, 2008).

Margaret McMillan believed that diet influenced both the physical and intellectual development of a child – "a hungry [malnourished] child cannot study" – that particular idea is relevant today. At the start of the 21st century we were most concerned that the diet of children in the UK was sufficient to meet their nutritional demands. Margaret's solution in 1906 was to lobby for the introduction of school meals; a hundred years later [2005] Jamie Oliver sought to improve the nutritional quality of school meals – the basic notion was common to both. More recently the Children's Society launched its 'Fair and Square' campaign [2012], aimed at inducing the Government to extend free school meals to all children living in poverty. The Society's report (Royston, Rodrigues and Hounsell, 2012) argued that eating a nutritious meal at lunchtime has important health and educational benefits for children; can improve their diet and increase their concentration during afternoon lessons and can potentially decrease health inequalities. Their argument echoes Margaret McMillan's in 1906.

If Margaret's successes at influencing policy are considered in isolation, without contemplating the means by which she achieved those ends, we would miss a valuable area of professional practice. Margaret was not by birth, position or gender an accepted policy-maker of her time. Margaret sought to develop her credibility by 'networking', debate and research. She actively published her ideas. She forged alliances and actively promoted her cause(s) and she sought to base her arguments on sound evidence, initiating her own trials where necessary. In modern 'management-speak' Margaret was adept at maximising her 'personal power' (Northouse, 2013). She portrayed herself as a leader for change; she emphasised her experience, education and expertise through her published writing and public speaking; she conveyed the impression that she had access to useful information through her research and contacts; and her high-profile supporters meant she would have been perceived as 'worthy' of a hearing. Her approach provides a model for any leader of change.

Conclusion

During Margaret's lifetime the notion of public health changed. It is widely agreed (Haynes, 2006) there were three phases: 1840 to c. 1880: new developments in sanitary legislation and engineering, belief in the 'miasma theory' and a laissez-faire attitude to private spaces; 1880 to 1910: an increased understanding of infection routes and bacteriology led to new infectious-disease policies; and c. 1910 onwards: a focus on personal health services, health education and individual hygiene, and so forth (Frazer, 1950). Margaret was in the vanguard for promoting new ideas – what is not clear is whether she was the originator or the conduit. It is not in doubt that Margaret instigated changes that resulted in improvements. She also had the concept that poor diet and environment affect the academic achievement of children, enshrined in British social policy, by successfully initiating the adoption of new areas of legislation (1906 and 1907 Acts).

This chapter has focused heavily on Margaret's political development, affiliations and contacts. This was deliberate; it was Margaret's ability to gain credibility and position herself so as to influence public policy that made her agitation for social reform effective. Her successes were lasting, but in the context of the limitations of the time, imposed by being female and yet seeking to influence policy, her successes were remarkable and her ability to forge alliances was fundamental to her success.

The determinants of health are better understood in the 21st century and are the foundation of current public health practice but it was people like Margaret McMillan, with both social conscience and the drive to alter things, that provided the early ideas and evidence on which we now build.

Bibliography

BJO, British Journal of Ophthalmology. (1941) Volume 25, pp. 592–593. in British Medical Journal 1941 *Obituary of Dr J Kerr, 5 October 1941.* http://bjo.bmj.com/ on March 14, 2016 Published by group.bmj.com; http://bjo.bmj.com/content/25/12/592.full.pdf

Cham, C. (1989) *Kropotkin: And the Rise of Revolutionary Anarchism, 1872–1886.* Cambridge University Press. https://books.google.co.uk/books?id=zk7wTSWT3JUC&printsec=frontcover&dq=Kropotkin&hl=en&sa=X&ved=0ahUKEwi85bu9y8rLAhVGsxQKHe41At8Q6AEIJTAA#v=onepage&q=Kropotkin&f=false accessed 17 March 2016.

Cole, G. D. H. (1905) *British Working Class Politics, 1832–1914.* London: George Routledge & Sons Ltd., p. 92.

Cresswell, D. (1948) *Margaret McMillan – a Memoir.* London: Hutchinson The writings of Margaret McMillan.

Duxbury-Neumann, S. (2015) *Little Germany: A History of Bradford's Germans.* Bradford: Amberley Publishing.

Education (Administrative Provisions) Act 1907. Chapter 43, 28th August 1907, London: HMSO.

Education (Provision of Meals) Act 1906. Chapter 57, 21st December 1906, London: HMSO. www.legislation.gov.uk/ukpga/1906/57/enacted accessed 1 April 2016.

Ewan, F., Innes, S., Reynolds, S., and Pipes, R. (eds.) (2006) *The Biographical Dictionary of Scottish Women: From the Earliest Times to 2004.* Edinburgh: Edinburgh University Press.

Frazer, W. M. (1950) *A History of English Public Health, 1834–1939.* London: Bailliere, Tindall & Cox, 1950, pp. vii–viii.

Gillard, D. (2003, June) *Food for Thought: Child Nutrition, the School Dinner and the Food Industry.* www.educationengland.org.uk/articles/22food.html accessed 14 March 2016.

Hansard (1924) *House of Commons Debate 10 July 1924,* Vol. 175 cc2488–9W.

Haworth, A., and Hayter, D. (2015) *Men Who Made Labour.* London: Taylor and Francis Publishers, p. 115.

Haynes, J. R. (2006) *Sanitary Ladies and Friendly Visitors: Women Public Health Officers in London, 1890–1930,* PhD Thesis University of London. http://eprints.ioe.ac.uk/19281/2/430969_Redacted.pdf accessed 4 March 2016.

Hughes, K. (1993) *Victorian Governess.* London: Hambledon Press.

Kier Hardy at Labour Representation Committee. (1900, 27 February) in *Report of the Labour Representation Conference 1900,* in *The Labour Party Foundation Conference and Annual Conference Reports 1900–1905* (Hammersmith Bookshop, London, 1967), 31.

Jarvis, P. (2013): *The McMillan Sisters and the 'Deptford Welfare Experiment'.* http://tactyc.org.uk/wp-content/uploads/2013/11/Reflection-Jarvis.pdfJHC, Journals of the House of Commons (1906) *House of Commons Debate 7 December 1906.* Volume 161, from 13th February 1906 to 21st December 1906. www.

publications.parliament.uk/pa/cm/cmjournal.htm and http://assets.parliament.uk/ Journals/HCJ_volume_161.pdf accessed 31 March 2016.

Jones, Peter d'Alroy. (1968) *The Christian Socialist Revival 1877–1914: Religion, Class, and Social Conscience in Late-Victorian England*. London: Oxford University Press.

Kropotkin, P. (1880) An Appeal to the Young: First appeared in French, 1880. "Aux Jeunes Gens". *Le Révolté*, June 25; July 10; August 7, 21. In English. http://dwardmac.pitzer.edu/Anarchist_Archives/kropotkin/appealtoyoung.html accessed 3 March 2016.

McMillan, M. (1896) *Child Labour and the Half-Time System*. Clarion Pamphlet No 15. London: Clarion Newspaper Co.

McMillan, M. (1899) *First Years of Childhood*. Manchester: Co-operative Newspaper Society (Read at the annual meeting of the Women's Co-operative Guild, Plymouth, July, 1899).

McMillan, M. (1900) *Early Childhood*. London: Swan Sonnenschein.

McMillan, M. (1903) *The Beginnings of Education*. London: Independent Labour Party (pamphlet).

McMillan, M. (1904) *Education Through the Imagination*. London: Swann Sonnenschein and Co. http://vufind.carli.illinois.edu/all/vf-nby/Record/hat_381883

McMillan, M. (1905) *The Economic Aspects of Child Labour and Education*. London: P. S. King.

McMillan, M. (1905) *Infant Mortality*. Pamphlet. London: Independent Labour Party.

McMillan, M. (1906) *Feeding London's School Children*. Pamphlet. Keithly: W. Yorkshire.

McMillan, M. (1907) After the Echoes of the Congress of School Hygiene. *The Labour Leader*, 30 August 1907.

McMillan, M. (1907) *Labour and Childhood*. London: Swan Sonnenschein. http://vufind.carli.illinois.edu/all/vf-nby/Record/hat_609540

McMillan, M. (1908) *The Mission of Children*. Pamphlet. London: Union of Ethical Societies.

McMillan, M. (1909) *The Bard at the Braes*. Pamphlet. London: Independent Labour Party. [fiction].

McMillan, M. (1909) *London's Children: How to Feed Them and How Not to Feed Them*. Pamphlet. London: Independent Labour Party.

McMillan, M. (1909) *New Life in Our Schools: The Clinic – What It Is and Is Not*. Pamphlet. London: Women's Co-Operative Guild.

McMillan, M. (1911) *The Child and the State*. Pamphlet. Manchester: National Labour Press. http://vufind.carli.illinois.edu/all/vf-nby/Record/hat_307353

McMillan, M. (1911, March) School Nursing in England. *The American Journal of Nursing*, 11(6), 459–464. www.jstor.org/stable/pdf/3405020.pdf?acceptTC=true

McMillan, M. (1912) *The School Clinic To-Day: Health Centres and What They Mean to the People*. Pamphlet. Manchester: National Labour Press.

McMillan, M. (1917) *The Camp School*. London: George Allen and Unwin Ltd.

McMillan, M. (1919) *The Future of Education Among Adolescents*. An address by Miss Margaret McMillan, delivered at the special educational session of the 51st annual co-operative congress, Carlisle, Tuesday, 10th June, 1919. Manchester, Co-operative Union.

McMillan, M. (1919) *The Nursery School*. London: J. M. Dent & Sons, Ltd. http://vufind.carli.illinois.edu/all/vf-nby/Record/3973180

McMillan, M. (1920) *What Is Democratic Education?* London: Workers' Educational Association. http://146.87.210.2/detail.aspx?parentpriref=

McMillan, M. (1924) *Education Through the Imagination*. New York: Appleton. http://vufind.carli.illinois.edu/all/vf-nby/Record/4950401 [educational psychology].

McMillan, M. (1924) *What the Open-Air Nursery School Is*. Pamphlet. London: Labour Party. http://146.87.210.2/detail.aspx?parentpriref=

McMillan, M. (1927) *The Life of Rachel McMillan*. London: J. M. Dent & Sons, Ltd.

McMillan, M. (1927) *NSA Memo*. Nursery School Association material from local groups deposited by Grace Owen @ British Library of Political and Economic Science. http://archives.lse.ac.uk/Record.aspx?src=CalmView.Catalog&id=BAECE%2f13%2f2&pos=1

McMillan, M. (1927–30) *Correspondence* prior to and following her resignation from the Nursery School Association committee, 29 November 1927–22 January 1930. http://archives.lse.ac.uk/Record.aspx?src=CalmView.Catalog&id=BAECE%2f13%2f8&pos=2

McMillan, M. (1928) *Nursery Schools and the Pre-School Child*. Pamphlet. London: Nursery School Association.

McMillan, M. (1930) *What the Open-Air Nursery School Is*. Pamphlet. London: Nursery School Association.

McMillan, M., et al. (1897) *Forecasts of the Coming Century – by a Decade of Writers*. Manchester: Labour Press.

McMillan, M., et al. (1907) *The Case for Women's Suffrage*. London: T. Fisher Urwin.

McMillan, M., et al. (1918) *Women and the Labour Party*. London: Headly.

McMillan, M., et al. (1920) *Nursery Schools, a Practical Handbook, and a Series of Appendixes Containing Official Reports and Recommendations*. London: J. Bale, Sons & Danielsson, Ltd. http://vufind.carli.illinois.edu/all/vf-nby/Record/11079544

Morris, W. (1888) *The Collected Letters of William Morris, Volume II, Part B: 1885–1888*, Princeton Legacy Library, Princeton University Press 2014, pp. 520; pp. 640; and pp. 657. https://books.google.co.uk/books?id=zfv_AwAAQBAJ&pg=PA640&lpg=PA640&dq=john+glasse+1887&source=bl&ots=4dTpgItApp&sig=o0QMfV-OjoUpKzT_KmJGs5XDtEY&hl=en&sa=X&ved=0ahUKEwiRteiWycrLAhXFtRQKHQ84Bv4Q6AEIITAB#v=onepage&q=john%20glasse%201887&f=false accessed 17 March 2016.

Northouse, P. G. (2013) *Leadership: Theory and Practice*. Los Angeles: Sage Publications.

Parkin, A. (2011) *The Annual Report of the Joint Director of Public Health (Bradford and Airedale) 2011/12*. NHS Airedale, Bradford and Leeds and City of Bradford MDC. www.bradford.nhs.uk/wp-content/uploads/2011/08/Public-health-annual-report-2012.pdf accessed 17 March 2016.

Royston, S., Rodrigues, L., and Hounsell, D. (2012) Fair and Square: A Policy Report on the Future of Free School Meals. *The Children's Society 2012*. www.childrenssociety.org.uk/sites/default/files/tcs/fair_and_square_policy_report_final.pdf downloaded 14 March 2016.

Smith, J. (1845) *Bradford Sanitary Committee, Report 1845*, Doc. Ref: 66D78, West Yorkshire Archive Service. Bradford. www.archives.wyjs.org.uk/

Steedman, C. (1990) *Childhood, Culture, and Class in Britain: Margaret McMillan 1860–1931*. New Brunswick, NJ: Rutgers University Press.

Steedman, C. (2004) *McMillan, Margaret (1860–1931), Socialist Propagandist and Educationist*. Oxford: Oxford University Press.

Stevinson, E. (1954) *Margaret McMillan: Prophet and Pioneer*. London: University of London Press.

UK Census 1901 at UK Census online. www.ukcensusonline.com/search/index.php?sn=McMillan&fn=Margaret&kw=&phonetic_mode=1&event=1901&source_title=Yorkshire+1901+Census&year=0&range=0&token=BMf_va6rjxSJDZ9E VeyOWT-iR6XVt-DbETaUnjem8WE&search=Search accessed 17 March 2016.

WHO. (2008) *Report of the WHO Global Commission on the Social Determinants of Health*. www.local.gov.uk/health/-journal_content/56/10180/3511260/ARTICLE#sthash.JrUndb11.dpuf accessed 14 March 2016.

9 George Cadbury and corporate social responsibility

Working conditions, housing, education and food policy

Zena Lynch and Surindar Dhesi

Introduction

Reports of poor employment and business practices in the news recently have brought the issues of Corporate Social Responsibility (CSR) and worker welfare directly to Parliament (House of Commons Business Innovation and Skills Committee, 2016). However, the idea of CSR and employer support for a resilient and productive workforce is not a new one; George Cadbury recognised the benefits and directly invested in employee and social welfare in the 1890s (Idowu, 2011). This chapter pursues the work of George Cadbury and discusses his legacy in modern business and social policy.

George Cadbury (1839–1922) was the son of John Cadbury who opened a grocers shop in Birmingham in 1824, which also sold cocoa and drinking chocolate (Cadbury, 2010). The chocolate business grew and in the 1890s a new factory was built and the Bournville Village developed on a greenfield site on the outskirts of the city (Bryson and Lowe, 2002).

George Cadbury was a Quaker and his religious beliefs heavily influenced his vision for workers welfare, both physical and spiritual. The Cadbury family faith prohibited exploitation of employees and others and 'promoted egalitarian, democratic relationships in the workplace' (Dellheim, 1987, p. 16). Cadbury considered himself accountable not only to God, but also to his family, employees, and the community (Parker, 2014). He also considered the acquisition of great wealth to be undesirable (Cadbury, 2010). However, he believed that taking care of workers was good business sense as it encouraged them to support the company in return (Idowu, 2011).

George Cadbury and his brother Richard pursued social welfare strategies and were well-known philanthropists. In addition to employee accommodation, they built a hospital, a convalescent home for children, and provided education for workers. However, this unusual approach and commitment to social justice led to political and industrial opposition (Parker, 2014).

Cadbury's legacy

Corporate social responsibility

CSR has been described as 'having fairness and morality in the conduct of an entity's dealing with all its stakeholders regardless of whether these are primary or secondary, internal or external' (Idowu, 2011, p. 152). Organisations committed to CSR have competitive advantages such as improved employee motivation and productivity, in addition to lower labour costs, improved health and safety, and lower premium rates from insurers and lenders (Parker, 2014).

CSR is an evolving concept and motivation is often related to company self-interest such as gaining permits and permissions more smoothly or easing customer relations (Frankental, 2011). In the Netherlands, it was found that intrinsic, moral motivations led to a greater commitment to CSR than where extrinsic, strategic motivations were more dominant (Ven van de and Graafland, 2006). Others have reported that where CSR comes into conflict with short-term financial objectives, the commitment can be abandoned (Parker, 2014).

The Cadbury brothers were not necessarily seeking to reduce costs or show a financial return on their social investment and their commitment went beyond their company walls into communities with promotion of the 'garden city movement' (Bryson and Lowe, 2002). They fostered the idea of 'cooperation through shared ideals' and believed this was reliant on the employment relationship (Fitzgerald, 1999 p. 168). They also focussed on producing good citizens as well as manufacturing quality products. The concept of shared ideals and CSR is interesting; it has been found that attitudes of employees to modern-day CSR are informed by a person's individual attitudes to society, as well as their perceptions of their employing company (Rodrigo and Arenas, 2008). CSR perceptions can be interest-based, and may vary relative to personal morals, values, and priorities (Green and Peloza, 2011). CSR can be a useful marketing tool and it has been observed that the 'genius of the Cadbury Brothers as developers, was their ability to build on open spaces for profit, while gaining reputations as guardians of the countryside' (Chance, 2007, p. 202).

Housing

George Cadbury was convinced that many social issues were connected to poor quality and overcrowded housing and was keen to build good accommodation at a reasonable expense. He explained his position in an interview stating 'We must destroy the slums of England or England will be destroyed by the slums. . . . We must not house our workers in a vile environment and expect their lives to be clean and blameless. We must do justice in the land' (Cadbury, 2010, p. 147). He strongly believed that living in poor environments resulted in 'diminished power to resist temptations' and this view has been supported with evidence in more contemporary work on the social determinants of health (Marmot, 2010).

After visiting people in the back streets of Birmingham, Cadbury concluded that providing access to open countryside and natural surroundings was necessary (Bailey and Bryson, 2006) and based on this principle, signed up to the 'garden city' movement promoted by Sir Ebenezer Howard (Miller, 2010). He created the Bournville estate in Birmingham (from 1893), which is described as a spacious and green place with low-density housing and large gardens for the cultivation of vegetables (Bryson and Lowe, 2002, Bailey and Bryson, 2006). Cadbury also became engaged with the aesthetic appearance of the housing and the arts generally and became a patron of the Arts and Crafts movement in Birmingham. His architect, William Harvey, set a tone of simplicity and modesty using brick and stone materials, which reflected his principles and identity (Bailey and Bryson, 2006). Cadbury's Quaker values also strongly influenced his attitude to the accumulation and inheritance of excessive wealth (Dellheim, 1987). By setting up the Bournville Village Trust in 1900, he removed himself from direct ownership of the village (Bryson and Lowe, 2002). This has been described by some as a move away from a paternal approach to housing (Bailey and Bryson, 2007), although others suggest that paternalistic control remained (Parker, 2014) (see Figure 9.1).

The investment in Bournville led to considerable improvements in public health. For the five years ending 1910, the death rate of 5.7 per 1,000 contrasted favourably to the national rate of 14.6 per 1,000 and infant mortality was substantially less than the national average. This carried on and by the early 1930s, the Bourneville adult and infant death rates were lower than the national averages (Parker, 2014).

The company town housing model has been described as a flawed model in that its sustainability depends on the success of the business; if the company fails, homes are also at risk. George Cadbury appeared to have had some foresight with respect to this issue, as he was very keen for Bourneville to be a mixed community, rather than just a 'workers village' and the numbers of residents working at the factory never rose above 50% (Jeffries, 2012). Today we also see the importance of housing provision to the jobs market; an imbalance can disrupt employees' home and work life which then affects employers (Walker, 1999). We also see that employment-related housing provision is more likely to be an individual 'fringe benefit', rather than part of a CSR or other corporate strategy (Doogan, 1996).

Figure 9.1 Bournville Village Green
Source: © Jill Stewart 2016.

Working conditions

Whilst Cadbury's policies in relation to his workforce are often lauded, closer inspection reveals them as somewhat mixed.

The factory move to Bournville is generally presented as a positive development for staff. However, at the time it was an inconvenience for many of the workers as it was some distance from the city where they lived and public transport was limited. Bournville Village accommodation was not affordable for the majority of factory employees and this necessitated the provision of the eating and other facilities so often admired (Bryson and Lowe, 2002). As was common at the time, Cadbury employed children from the age of around 14, and women lost their jobs when they married and were given a Bible, a flower, and a small monetary sum when departing (Dellheim, 1987). Within the factory and its grounds men and women were strictly segregated (Rowlinson and Hassard, 1993).

Cadbury was particularly innovative by providing doctors and dentists in the factory, and also in setting out procedures for the prevention and investigation of accidents and fires. He also introduced sporting facilities (Chance, 2007), a pension scheme for workers (Parker, 2014), and paid higher wages than were usual in the area (Chance, 2007).

In terms of the cocoa-growing workforce, working conditions were quite different as plantations were associated with slave labour. The Cadbury company (then run by William Cadbury) (Higgs, 2014) was heavily criticised for its hypocrisy in buying cocoa from slave plantations on the Portuguese islands of São Tomé and Principe in West Africa (Dellheim, 1987). A comparison of the slave and Bourneville factory conditions reported in *The Standard* newspaper led to a court case in 1909, which the company won, but were awarded one farthing as damages, a tiny sum at the time (John, 2002). This was particularly embarrassing, since the older generation of Cadburys had been 'involved in countless ant-slavery ventures' (Hasian, 2008, p. 250).

In 2008, it was reported that the Ivory Coast which produces around half of the world cocoa crop was using 'hundreds of thousands of children' in modern slavery on the plantations (Hasian, 2008, p. 268).

Education

George Cadbury valued education and supported programmes for the workforce and also the local community. He was instrumental in the development of Adult Schools and also showed his commitment by personally teaching in them (Parker, 2014).

The children that the Cadburys employed (children were employed extensively in factories at this time) (Tuttle, 2001) were expected to finish work an hour early twice per week to attend evening classes. For those wishing to pursue their education, day classes together with scholarships were also offered (Parker, 2014). There was a separate Works Education Committee set up to support general education including 'practical living skills, physical education, commercial training, apprenticeship training and technical trade training' (Parker, 2014, p. 646).

Considering the picture today, education investment by business seems to be mainly the realm of the multinationals, with the top 10 Global 500 spenders, from 2011 to 2013 contributing to nearly half (42%) of the total education-related corporate spending (Pota, 2015). Looking further afield for contemporary comparisons, although trades unions (Ball, 2011) and co-operatives (Wadsworth, 2012) are known to invest in education of their members, these are not employer–employee relationships and employer investment in education not directly related to the work activity remains unusual. An investigation by Arnould, Alejandro and Ball (2009) also found that although co-operative involvement affects the likelihood of children being educated, the effects on educational and health outcomes are uneven.

Product quality

Food adulteration has long been an issue. In Cadbury's time it was common for English cocoa to be adulterated with starch such as potato flour or sago to mask the excess cocoa butter. The cocoa drink, as described by George Cadbury himself, was a 'comforting gruel' (Dellheim, 1987, p. 17). However, the quality was poor.

Cadbury introduced a new cocoa press, which greatly improved the quality, by removing the excess cocoa butter and thus removing the need to add starch. This enabled the product to be marketed as 'Absolutely pure . . . therefore Best' (Dellheim, 1987, p. 17), a claim which was supported by medical journals of the time and also rabbis. This distinction from competitors and the promotion of the product as a cheap bedtime drink (Higgs, 2014) helped to boost sales and save the struggling business (Parker, 2014).

The move of the factory to the green area of Bournville also enabled marketing claims to be made about the purity and wholesomeness of the product and the process carried out in the 'factory in a garden' (Bryson and Lowe, 2002, p. 26). Today, we also commonly see claims of 'natural' and 'pure' being effectively used for marketing foods (Farris, 2010), product quality being tightly linked as observable proof of a company's commitment to its core values. However, concerns from sceptical consumers around the meaning of certain terms has led to the production of guidance on their use (Food Standards Agency, 2008).

Paternalism

As we have shown throughout this chapter, Cadbury's concern for the well-being of his workforce and others often tipped into paternalism. A further example was the control exerted on employees' outside activities, particularly with regard to intemperance. Cadbury did not permit any public houses in Bournville (Dellheim, 1987) where new residents were strongly advised to avoid tobacco, alcohol, and pork (Bailey and Bryson, 2006). Letters were also sent to residents who failed to maintain their gardens (Bryson and Lowe, 2002).

An incident was reported where an employee received a letter relating to the unacceptability of drunkenness at a station on a Saturday evening. There was a further report where an employee was angered when prevented from playing the piano in public houses when threatened with loss of his position in the factory. It also appeared that support was made available only to 'deserving citizens who understood the benefits of temperance, thrift and hard work' (Bailey and Bryson, 2006, p. 190).

In many respects these attitudes have endured and there has been a resurgence in 'conditionality' for welfare assistance in the UK in recent years, including measures aimed at changing behaviour, and as with Cadbury's sanctions these have been reported to have led to hostility to authority amongst some of those affected (Watts et al., 2014).

Conclusion

We have seen that George Cadbury's actions and legacy are rather more nuanced than is often presented. Employment law was undeveloped during the 18th and 19th centuries and employers assumed an automatic right to dictate terms of employment (Parker, 2014). This is reflected in some of the paternalistic actions taken against employees not complying with the required 'standard', even outside the workplace. As with many modern businesses committed to CSR, Cadbury was concerned to present his enterprise and products in the most 'wholesome' light.

Viewing CSR from the historical perspective, Cadbury's legacy highlights how societal values at a particular time influence company policy. The distinction between how money is made and how it is spent has also shifted over time. For example it has been argued that without profit George Cadbury would not have been in a position to make social investments in Bournville (Rowlinson, 1998), whereas today CSR would extend to consideration of the supply-chain. There is juxtaposition between doing social good and exploiting power to make decisions over people's lives and restricting their autonomy. Victorian value judgements as to whether individuals are 'deserving' of help remain in contemporary society.

George Cadbury was nevertheless a great pioneer in industrial philanthropy, standing apart from many of his peers with a clear vision and principles. The complex and challenging issues he addressed in the 18th and 19th centuries, including working and living conditions, child labour, education, product quality remain of concern. When he died, 16,000 people attended his memorial (Dellheim, 1987) and is difficult to imagine his present-day equivalents inspiring such regard.

Bibliography

Arnould, E., Alejandro, P., and Ball, D. (2009) Does Fair Trade Deliver on Its Core Value Proposition? Effects on Income, Educational Attainment, and Health in Three Countries. *Journal of Public Policy & Marketing*, 28, 186–201.

Bailey, A. R., and Bryson, J. R. (2006) Stories of suburbia (Bournville, UK): From Planning to People Tales. *Social & Cultural Geography*, 7, 179–198.

Bailey, A. R., and Bryson, J. R. (2007) A Quaker Experiment in Town Planning: George Cadbury and the Construction of Bournville Model Village. *Quaker Studies*, 11.

Ball, M. (2011) Learning, Labour and Union Learning Representatives: Promoting Workplace Learning. *Studies in the Education of Adults*, 43, 50–60.

Bryson, J. R., and Lowe, P. A. (2002) Story-Telling and History Construction: Rereading George Cadbury's Bournville Model Village. *Journal of Historical Geography*, 28, 21–41.

Cadbury, D. (2010) *Chocolate Wars: The 150-Year Rivalry Between the World's Greatest Chocolate Makers*. New York: PublicAffairs.

Chance, H. (2007) The Angel in the Garden Suburb: Arcadian Allegory in the ('Girls' Grounds' at the Cadbury Factory, Bournville, England, 1880–1930. *Studies in the History of Gardens & Designed Landscapes*, 27, 197–216.

Dellheim, C. (1987) The Creation of a Company Culture: Cadburys, 1861–1931. *The American Historical Review*, 92, 13–44.

Doogan, K. (1996) Labour Mobility and the Changing Housing Market. *Urban Studies*, 33, 199–221.

Farris, A. L. (2010) The Natural Aversion: The FDA's Reluctance to Define a Leading Food-Industry Marketing Claim, and the Pressing Need for a Workable Rule. *Food & Drug Law Journal*, 65, 403–424.

Fitzgerald, R. (1999) Employment Relations and Industrial Welfare in Britain: Business Ethics Versus Labor Markets. *Business and Economic History*, 28, 167–179.

Food Standards Agency. (2008) *Criteria for the Use of the Terms Fresh, Pure, Natural etc. in Food Labelling Food Standards Agency*.

Frankental, P. (2011) Corporate Social Responsibility – a PR Invention? *Corporate Communications: An International Journal*, 6, 18–23.

Green, T., and Peloza, J. (2011) How Does Corporate Social Responsibility Create Value for Consumers? *Journal of Consumer Marketing*, 28, 48–56.

Hasian, M. (2008) Critical Memories of Crafted Virtues: The Cadbury Chocolate Scandals, Mediated Reputations, and Modern Globalized Slavery. *Journal of Communication Inquiry*, 32, 249–270.

Higgs, C. (2014) Happiness and Work: Portuguese Peasants, British Laborers, African Contract Workers, and the Case of São Tomé and Príncipe, 1901–1909*. *International Labor and Working-Class History*, 86, 55–71.

House of Commons Business Innovation and Skills Committee (2016) *Employment Practices at Sports Direct*. www.parliament.co.uk.

Idowu, S. O. (2011) An Exploratory Study of the Historical Landscape of Corporate Social Responsibility in the UK. *Corporate Governance*, 11, 149–160.

Jeffries, D. (2012) Home From Work: The Company Town Housing Model. *The Guardian Housing Network*. www.theguardian.com/housing-network/2012/feb/24/home-work-company-town-model

John, A. V. (2002) A New Slavery? *History Today*, 52, 34–35.

Miller, M. (2010) *English Garden Cities: An Introduction*. Swindon: English Heritage.

Parker, L. D. (2014) Corporate Social Accountability Through Action: Contemporary Insights from British Industrial Pioneers. *Accounting, Organizations and Society*, 39, 632–659.

Pota, V. (2015) Companies Are Spending Too Little on Education – and in the Wrong Places. *The Guardian*. www.theguardian.com/sustainable-business/2015/jan/14/companies-business-education-health-jobs-employees-careers accessed 6 October 2016.

Rodrigo, P., and Arenas, D. (2008) Do Employees Care About CSR Programs? A Typology of Employees According to Their Attitudes. *Journal of Business Ethics*, 83, 265–283.

Rowlinson, M. (1998) Quaker Employers. *Historical Studies in Industrial Relations*, 6, 16–198.

Rowlinson, M., and Hassard, J. (1993) The Invention of Corporate Culture: A History of the Histories of Cadbury. *Human Relations*, 46, 299.

Tuttle, C. (2001) Child Labor During the British Industrial Revolution [Online]. *EH.Net Encyclopedia.* http://eh.net/encyclopedia/child-labor-during-the-british-industrial-revolution/

Ven Van de, B., and Graafland, J. J. (2006) *Strategic and Moral Motivation for Corporate Social Responsibility*. St. Louis: Federal Reserve Bank of St Louis.

Wadsworth, J. (2012) Study Raises Important Questions About Co-Op Education Efforts. *Rural Cooperatives*, 79, 25–26.

Walker, R. M. (1999) The Housing Problems of Employees: Housing Markets, Policy Issues and Responses in England. *Netherlands Journal of Housing and the Built Environment*, 14, 143–161.

Watts, B., Fitzpatrick, S., Bramley, G., and Watkins, D. (2014) *Welfare Sanctions and Conditionality in the UK*. York: Joseph Rowntree Foundation. Available at https://www.jrf.org.uk/report/welfare-sanctions-and-conditionality-uk

10 Charles Booth's inquiry

Poverty, poor housing and legacies for environmental health

Matthew Clough

Introduction

Charles Booth's 1902 Inquiry 'Life and Labour of the People of London' was his attempt to understand 19th-century housing in how we understand poverty. A synopsis of the Inquiry's methods and findings highlights key facts and draws links to contemporary housing issues with specific reference to the work of the Environmental Health Practitioner (EHP). The discussion covers the indirect effects of the Toynbee Hall community to address poverty, Booth's indirect legacy to spatial planning and scientific research, which all feed into both general environmental health practice and healthy places to live. Lastly the chapter explores early regeneration, looking at the East London Boundary Estate renewal scheme and the lessons in securing healthier housing.

Charles Booth: inquiry and motivations

Charles Booth (Figure 10.1) was a successful businessman of Alfred Booth & Company whose enterprise was primarily shipping goods (LSE, 2016). From Liverpool, Booth travelled widely, although eventually he moved to live in London. It's thought the influence of his business networking of associates, the formation of friendships with people such as Octavia Hill, Samuel and Henrietta Barnett, combined with his eye for detail and organisational skills led to his study of London's poor. Booth's study of the poor covered the period of the late 19th and early 20th centuries. He also had interests in political debating and campaigned for old age pensions.

Booth's survey into poverty of the time is thought to be the earliest scientific study of its kind. Data were collected from school board visitors and cross-referenced with interviews of prominent philanthropists, police and other influential people (Englander, 1998, p. 57). Facts, figures and detailed descriptions into the housing, cost of rents, employment and religious and charitable works across East and West London were published into three

Figure 10.1 Portrait of Charles Booth

Source: Wellcome Library, London.

series across 17 volumes. The Inquiry also drew on census data of 1891 (Booth, 1902b). Overall the level of detail and vivid descriptions make the published works an interesting and informative read.

The Inquiry findings give insight into Booth's analytical mind; he published tables with the following headings: gender, birthplace, industrial status and levels of overcrowding showing populations within each distinct area (Booth, 1902a). Moreover Booth includes descriptive notes which are consistent area by area, using the same headings covering housing, rents, trades and employments, market places, public houses, health and locomotion. Indeed by revisiting the same areas over 10 years later, the Inquiry was able to study the effect of changes over time in key determinants.

By 1889 Booth was able to use his classification system (Table 10.1) to plot data on maps to see geographical trends in poverty; his maps show

Table 10.1 Booth's Social Classification system

Booth Classification	Description of Class
A (Black)	The lowest class which consists of some occasional labourers, street sellers, loafers, criminals and semi-criminals. Their life is the life of savages, with vicissitudes of extreme hardship and their only luxury is drink.
B (Dark Blue)	Casual earnings, very poor. The labourers do not get as much as three days' work a week, but it is doubtful if many could or would work full time for long together if they had the opportunity. Class B is not one in which men are born and live and die so much as a deposit of those who from mental, moral and physical reasons are incapable of better work.
C (Light Blue)	Intermittent earning. 18s to 21s per week for a moderate family. The victims of competition and on them falls with particular severity the weight of recurrent depressions of trade. Labourers, poorer artisans and street sellers. This irregularity of employment may show itself in the week or in the year: stevedores and waterside porters may secure only one of two days' work in a week, whereas labourers in the building trades may get only eight or nine months in a year.
D (Purple)	Small regular earnings. poor, regular earnings. Factory, dock and warehouse labourers, carmen, messengers and porters. Of the whole section none can be said to rise above poverty, nor are many to be classed as very poor. As a general rule they have a hard struggle to make ends meet, but they are, as a body, decent steady men, paying their way and bringing up their children respectably.

Booth Classification	Description of Class
E (Light Pink)	Regular standard earnings, 22s to 30s per week for regular work, fairly comfortable. As a rule the wives do not work, but the children do: the boys commonly following the father, the girls taking local trades or going out to service.
F (Dark Pink)	Higher-class labour and the best paid of the artisans. Earnings exceed 30s per week. Foremen are included, city warehousemen of the better class and first hand lighter-men; they are usually paid for responsibility and are men of good character and much intelligence.
G (Red)	Lower middle class. Shopkeepers and small employers, clerks and subordinate professional men. A hardworking sober, energetic class.
H (Yellow)	Upper middle class, servant-keeping class.

Source: Reproduced and adapted with permission from the London School of Economics Charles Booth online archive.

concentrations of people in the lowest classes away from the main through routes around East London and close to the docks. Out of Booth's work came the poverty line measure. At that time he established a family needed approximately 21 shillings a week to subsist (Hatchett et al., 2012, p. 20). Taking account of inflation, the Bank of England estimates this equates to £115.00 in 2015, a figure very close to the current couple's rate for Job Seekers Allowance (Bank of England, 2016).

The inquiry showed a strong link between level of household income and quality of the housing people occupied. It is these parallels, which are discussed further in the following sections.

Poverty from Booth's perspective and today

In understanding poverty there is a need to have robust and reliable data. Booth surprised himself in that 30% of the population lived in dire poverty. Originally he thought the figure was around 25% (LSE, 2016). In the approach to getting this data Booth aimed to be sympathetic to his subject's situation and be careful just to be an observer, so it could be argued this survey was an early form of informed consent (responsibility must be given to the observed individual to make the decision to participate), a key principle of investigative study for those involved in research. His published work shows he was aware of bias and his non-critical approach to the subjects studied facilitated greater access. His qualitative research was guided by what he saw with his own eyes and interviews over a period of time (Booth, 1902a, p. 5).

Booth was critical of the inertia of elected officials, writing about the Poplar, Bromley and Bow vestries (local government of the day):

They were cautious slow going people and their statutory duties were very light.

(Booth, 1902a, p. 60)

It was not until the legislative changes of 1894, which abolished the qualification requirement for persons taking up public office of owning property that there was much public health–related municipal action. Further change came with the London Government Act 1899, in which the overall effect transferred much power to the working classes. Working-class people become poor law guardians and distributed relief. Later local government action in the locality led to the creation of baths and wash houses, libraries and improved sanitation with associated health benefits (Booth, 1902a, pp. 61–61).

The author notes that overcrowded housing in London is not a new phenomenon. Booth's study of poverty highlighted in detail the insanitary impacts of overcrowded housing in the 19th century (Spicker, 1989). People lived in overcrowded accommodation because it was all they could afford. Organisations such as the Joseph Rowntree Foundation (JRF, 2016) published a 'London Mayoral Manifesto for 2016' lobbying for policies, including more affordable accommodation for Londoners, an issue yet to be fully addressed.

Booth has a legacy into town planning and urban layout; his published Inquiry data are so comprehensive they have been used by researchers such as Professor Laura Vaughan to investigate the relationship between spatial form and why poverty has persisted in some areas. Using the technique space syntax analysis, Vaughan (2007) has compared poverty in Booth's time with the late 20th century, finding correspondence between poverty and spatial separation, which tended to persist in isolated communities, thus highlighting the need for social integration. Others such as Spicker (1989) also found urban layout has impact, for example pockets of poverty are found in areas isolated by railway lines. It is important not to forget these key issues to which EHPs often have opportunity to contribute through consultation; failure to take them into account could have unforeseen health effects.

Impact of Booth's findings on housing standards

Data from Booth's Inquiry clearly influenced the political debate on sanitation, Poor Laws and Victorian/Edwardian attitudes to poverty. The Artisans and Labourers Dwelling Act 1875 (Cross Act) and various Torrens Acts implemented to enable Local Authorities to clear slum areas had failed to make significant impact in improving standards in parts of East London

(Wise, 2009). There is support to the argument of a paradigm shift over a period of time in political thinking on poverty relief and the raising of housing standards. For example The Public Health (London) Act, 1891 was a clear legislative attempt to bring a more level playing field to dealing with sanitary deficiencies (Great Britain, 1891).

While elsewhere the public could invoke the sanitary code with provisions contained in Public Health Act 1875, in London it was only by 1891 that sanitary inspectors now had an effective legislative tool for ensuring housing developers included water closets and ash pits within new housing developments. Notices could be served and fines administered for landlords who failed to comply with direction (Great Britain, 1891). Key contemporary legislation such as the Environmental Protection Act 1990 has a similar layout and approach, highlighting the roots of this legislation in tackling public nuisances (Bell and McGillivray, 2006, p. 404).

On a localised level pioneers such as Canon Samuel Barnett moved to the East End of London so as to make a difference for the poor. Barnett set up Toynbee Hall community in Aldgate to create a place to enable the development of practical ideas and solutions to alleviate and remediate poverty. Barnett had a long-term commitment to the area, setting up social housing and the Toynbee community ran cultural events (Till, 2013). Indeed Barnett enabled key people to join Booth's research team and the two men were friends, both being critical of poor laws while sharing similar ideas on how to tackle social problems.

Toynbee Hall's 'Poor Man's Lawyer' enabled the East End population to access housing advice, clearly a forerunner of legal advice clinics of today. This is often a service the modern EHP needs to direct beleaguered tenants towards, when landlords are imposing unfair and unlawful tenancy terms and conditions. Toynbee Hall continues to undertake research, influence thinking and lobby for policy to find long-term solutions for dealing with poverty. Around 42% of children living in Tower Hamlets today live in poverty (Till, 2013).

Booth and the Boundary Street regeneration

In looking at early regeneration the author explores the London County Council's (LCC) scheme to remove the 'Old Nichol' slums, an area sandwiched between Shoreditch and Bethnal Green (Figure 10.2). Booth comments starkly on the Old Nichol:

> A district of almost solid poverty and low life, in which the houses were as broken down and deplorable as their unfortunate inhabitants.
>
> (Booth, 1902b, p. 68)

The streets were given classifications of black or dark blue where there was severe overcrowding. Homes were damp, dark, had little ventilation or sanitation, general decay and dirt everywhere (BBC, 2012). Foul trades and death added to make conditions very unpleasant (Wise, 2009, p. 7).

The publishing of Arthur Morrison's *A Child of the Jago* that depicted a deplorable life with few opportunities except for crime aided an awareness campaign to the plight of slum dwellers led by local clergyman Rev Osborne Jay (Wise, 2009, p. 226). This literary work along with the observations of significant others such as vestry Medical Officer Dr George Bate brought the Nichol to the attention of Government. Bate highlighted 43% of Nichol homes were beyond redemption in making fit for human habitation.

The setting up of the LCC in 1888 led to key political change, ultimately the driver of regeneration in the Nichol (Wise, 2009, pp. 339–347). LCC took over responsibility for housing from the Metropolitan Board of Works. The housing committee persuaded the home secretary to grant permission for a demolition and rebuilding. The scheme was beset with mounting costs, overly generous compulsory purchase orders to former slum landlords, tenant compensation for displacement, non-payment of rents from those who remained and delays which included dealing with squatters.

Indeed Booth (1902b, pp. 71–72) highlights the need for great care in undertaking major regeneration. Jack London states in his *The People of the Abyss* the new estate was a healthier place to live, for higher classes of workmen and crafts people (Figure 10.2). The slum people drifted elsewhere, which resulted in the area becoming gentrified (London, 1903, p. 31).

The involvement of residents in housing renewal is a fairly recent policy move; however, the potential for improvement in health outcomes is significant (Anderson and Barclay, 2003). Within the context of a holistic approach to avoid past failings, environment health plays a key part in area renewal to address poor housing. The 'Commission on Housing Renewal and Public Health', a group of leading experts, many with EHP backgrounds, have put together detailed good practice guidelines (CIEH, 2007). The Commission puts together the arguments for and against regeneration and so the evidence needs to be weighed up carefully.

Conclusions

Booth the entrepreneur showed it was possible with appropriate skills, resources and well-connected contacts to undertake a major study of the social hierarchy and levels of poverty throughout London. The inquiry maps graphically show the link between low levels of income and low housing standards. Booth commentates on the debate to alleviate poverty, through

Figure 10.2 Shiplake House on Arnold Circus
Source: Author's own photograph.

reorganisation of local government and reform of legislation, of which both have a legacy to contemporary environmental health.

The wider impact of organisations such as the Joseph Rowntree Foundation and others, including Samuel Barnett, may not have been as significant without Booth; their collective approach clearly has an indirect legacy to dealing with housing issues. Booth's Inquiry coverage into early regeneration identified the political circumstances that brought about early slum clearance while also highlighting the need to have clearly defined aims and objectives for such major projects. These key principles should not be lost, if the lessons of the past are not to be repeated.

Acknowledgements

I wish to thank Tina Garrity, librarian at Chadwick Court and Jill Stewart at Greenwich University for her helpful guidance, loan of some reference material and support in writing this chapter.

94   *Matthew Clough*

Bibliography

Anderson, I., and Barclay, A. (2003) Housing and Health. In: *Public Health in Practice*, ed. A. Watterson. Basingstoke: Palgrave Macmillan, pp. 158–181.

Bank of England (2016) *Bank of England Inflation Calculator*. www.bankofeng land.co.uk/education/Pages/resources/inflationtools/calculator/flash/default.aspx accessed 24 June 2016.

BBC (2012) The Secret History of Our streets 2012 Season 1 Episode 6 [online]. www.youtube.com/watch?v=bBU-ctynq7o#t=28.1867872 accessed 7 July 2016.

Bell, S., and McGillivray, D. (2006) *Environmental Law* (6th ed.). Oxford: OUP.

Booth, C. (1902a) *Life and Labour of the People of London: Third Series Religious Influences*. London North of the Thames: The Outer Ring. London: Macmillan (Volume 1).

Booth, C. (1902b) *Life and Labour of the People of London: Third Series Religious Influences*. London North of the Thames: The Inner Ring. London: Macmillan (Volume 2).

CIEH (2007) Commission on Housing Renewal and Public Health – Final Report www.cieh.org/policy/commission_housing_renewal.html accessed 7 July 2016.

Englander, D. (1998) *Poverty and Poor Law Reform in 19th Century Britain, 1834–1914 From Chadwick to Booth*. London: Longman Pearson Education.

Great, B. (1891) *The Public Health (London) Act 1891 With an Introduction, Notes and an Index*. https://archive.org/details/publichealthlond00greaiala accessed 24 June 2016.

Gov.uk (2016) *Job Seekers Allowance (JSA)*. www.gov.uk/jobseekers-allowance/overview accessed 24 June 2016.

Hatchett, W., Spear, S., Stewart, J., Stewart, J., Greenwell, A., and Clapham, D. (2012) *The Stuff of Life Public Health in Edwardian Britain*. London: CIEH.

Joseph Rowntree Foundation. (2016) *A London Without Poverty: The Joseph Rowntree Foundation's Manifesto Briefing for the London Mayoral Election 2016*. www.jrf.org.uk/report/london-without-poverty accessed 29 June 2016.

London, J. (1903) *The People of the Abyss*. London: Journeyman Press.

London School of Economics & Political Science. (2016) *Charles Booth Online Archive, Charles Booth and the Survey Into Life and Labour in London (1886–1903)*. Maps Descriptive of London Poverty. http://booth.lse.ac.uk/static/a/4.html accessed 19 June 2016.

LSE. (2016) *Charles Booth Online Archive, Charles Booth and the Survey Into Life and Labour in London (1886–1903)*. http://booth.lse.ac.uk/accessed 7 July 2016.

Spicker, P. (1989) Charles Booth: Housing and Poverty in Victorian London. Published as Victorian values. *Roof*, 1989, pp. 14, 38–40; and subsequently reprinted in C. Grant (ed.) *Built to Last?*, Roof 1992, pp. 8–14 and J. Goodwin and C. Grant (eds.) *Built to Last*, Roof 1997.

Till, J. (2013) *Icons of Toynbee Hall Samuel Barnett*. www.toynbeehall.org.uk/our-history accessed 4 July 2016.

Vaughan, L. (2007) *The Spatial Form of Poverty in Charles Booth's London*. http://discovery.ucl.ac.uk/3273/1/3273.pdf accessed 7 July 2016.

Wellcome Library, London. Wellcome Images images@wellcome.ac.uk http://well comeimages.org Portrait of Charles Booth, Social Reformer after: Russell and Sons Publishers. Copyrighted work available under Creative Commons Attribution only licence CC BY 4.0 http://creativecommons.org/licenses/by/4.0/

Wise, S. (2009) *The Blackest Streets*. London: Vintage.

Further reading and websites

Bullman, J., and Hegarty, N. (2012) *The Secret History of Our Streets*. London, BBC: London. Book published with the BBC series charting the changes to streets mapped by Booth.

Jackson, L. (2014) *Dirty Old London, the Victorian Fight Against Filth*. New Haven and London: Yale University Press. Insight into hygiene challenges in Victorian London.

Museum of London, Charles Booth Maps.

www.museumoflondon.org.uk/

The Geffrye Museum covers the history of the home.

www.geffrye-museum.org.uk/

11 Christopher Addison

Health visionary, man of war, Parliamentarian and practical pioneer

William Hatchett

Introduction

There aren't many books devoted to Christopher Addison, the UK's first health minister. There are two biographies at the time of writing – that by Kenneth and Jane Morgan proved the most useful to this writer, with its frequent references to the politician's papers. Because of his unassuming personality, Addison has not received the credit that is his due; he does not gain a prominent place in other politicians' memoires, even if they knew him intimately. Perhaps aptly, he is best known for a major piece of legislation that bears his name, the Housing and Town Planning Act of 1919, which is often referred to as 'the Addison Act'. Gratifyingly, two streets of inter-war council housing are also named after him near his adopted town, Watford.

Born in Hogsthorpe in Lincolnshire, the youngest son of a substantial tenant farmer, who was a 'true blue' Conservative, Addison was sent to private school in Harrogate. Possessing a logical mind, a strong stomach and an appetite for science, he excelled at medicine, specifically anatomy (his understanding of complex systems and how they worked would later serve him extremely well in his political career). Addison became a fellow of the Royal College of Surgeons in 1895 and could easily have pursued a distinguished medical career. He is still known by anatomists for his discovery of 'Addison's planes', which are delineations of the stomach (Morgan and Morgan, 1980). That he chose insecure politics over secure and well-paid medicine indicates the strength of his commitment to public health reform. He became baron in 1937 and was elevated to viscount and leader of the House of Lords in 1945.

Contact with the poor in medical practice, for example working as a doctor at the Banstead Asylum, had stirred his social conscience. This led him to contest and win a Liberal Parliamentary seat in the working-class London district of Shoreditch at the age of 41, where he enthusiastically represented hawkers, costermongers and postal porters (Morgan and Morgan, 1980).

Entering the House of Commons in 1910, Dr Addison immediately became an advocate of the Liberals' radical political programmes in children's and maternal welfare and replacing poor law welfare provision with state benefits. Herbert Asquith, a remote, autocratic figure, still led the party. But Addison hitched his wagon to the Liberals' rising star, the then chancellor, Lloyd George, who would prove his mettle in the First World War and take over leadership of his party, and the war, in 1916, in a coalition with the Conservatives.

Addison wasn't a showy figure, a stump orator or a scintillating Parliamentary performer. He wasn't a great writer – his books including, *The Betrayal of the Slums*, 1922 (Addison, 1922), *Politics from Within*, 1923 and *Practical Socialism*, 1926, while at times passionate, are solid assemblages of pertinent facts – well-made, but not exciting. Whether participating in a debate or working on new legislation, he was thorough and logical and had a meticulous eye for detail.

His cheerful, positive manner must have made him popular, because, even though he was a tad colourless he was highly 'clubbable'. Most people liked him, with the exception of Neville Chamberlain, who called him a 'miserable cur' (Morgan and Morgan, 1980). But that did not stop Chamberlain from granting Addison a peerage in 1937. Addison and Lloyd George's working partnership was unlikely – the charismatic womaniser and the dry medic – but it worked, as did his later pairing with Clement Attlee (they were known to insiders as 'Clem and Chris'). His allegiances often crossed party lines. Perhaps because he came from a Presbyterian background, he was an unlikely but close friend to the Unionist politician, Sir Edward Carson, in his early period as an MP.

For his entire life (to the disapproval of some fellow Socialists) he maintained civil contact with Sir Winston Churchill, who was immensely grateful for Addison's work on Liberal social reform. Another friend and political ally was the millionaire, Unionist, 'pure food' campaigner and owner of *The Observer* newspaper, Waldorf Astor. With the flamboyantly homosexual and 'high church' Labour MP, Tom Driberg, Addison visited the newly liberated Buchenwald concentration camp with a group of Parliamentarians in 1945. He and Driberg were, reportedly, the only delegation members who viewed the horrors without 'breaking down' (Morgan and Morgan, 1980).

Crucially, Addison was Lloyd George's dependable formulator and ally as the key National Insurance Act of 1911 was nursed into life; he later served as Ramsey MacDonald's lieutenant and aide-de-camp (MacDonald was Labour's 'lost leader' in the disastrous and short-lived 1930 National Government). In the last phase of his career, which was almost as long, in total, as Churchill's, Addison served as Clement Attlee's most important 'fixer' in the House of Lords. His subtle skills and lifelong contacts were crucial to pushing the Socialist legislation of the 1945–51 government through a chamber with a large Conservative majority (including nationalisation of the mines and iron and steel industries as well as Labour's all-important health and national insurance legislation).

Because he was so diligent in his Parliamentary duties, his extant papers and dairies, which are well catalogued and held in the Bodleian Library, are of massive extent. Amazingly, his four decades of contributions to both houses of Parliament can be consulted free online, but they don't contain many fireworks. Even Addison's 'big moments' – such as the speech announcing his high-profile resignation from Lloyd George's coalition government in July 1921 – tend to be damp squibs. Addison was most prolific as a writer between 1922, when he lost his Liberal Hoxton seat, and 1929, when he was elected as a Labour MP for Swindon (having changed his party allegiance and failed to be elected in Hammersmith).

Aside from his full-length works, his chapters and pamphlets for the Socialist League and the Left Book Club from this time are mainly devoted to food supply and agriculture, drawing on his agricultural roots in rural Lincolnshire. These writings were a good preparation for his time as an agricultural adviser, then minister for Ramsey MacDonald, a position he achieved in 1930, before Labour lost the election and he again lost his seat (a repeat performance of the reversal of 1922).

You could say that Addison was simply one of those anonymous, backroom figures who make the wheels of government turn. But that would be to under-estimate, massively, his major contribution to UK public health. He applied his cool physician's eye, his industriousness and his skill at maintaining friendships (even he and Lloyd George eventually made up) to a political career which, while lacking in visibility, was far more productive than that of the huge majority of Parliamentarians.

Part of his career's effectiveness is due to its amazing longevity – 41 years, of which only 12 were spent out of office. Addison was already middle-aged when he became an MP, leaving behind a substantial medical reputation; he was aged 50 when he became health minister in 1919 and, by the standards of the time, would have been judged an old man in the 1920s, when, having lost his seat, he occupied his own version of Churchill's political wilderness (unlike Churchill, his political views shifted markedly to the left). He was 70 (5 years older than Churchill) at the outbreak of the Second World War and one of a few public servants whose experience of the previous war (including the state's adoption of unprecedented centralised powers) would be immensely valuable in the coming conflagration.

Addison was a transitional figure from the Victorian era and, also, a man of war. It was war which gave him control of the levers of public health and taught him how to use them. His legacies are those things that all of us from the baby boom generation have benefitted from (and are now steadily losing): state benefits for the sick and unemployed, including a 'safety net' which is not stigmatising or punitive (unlike Edwin Chadwick's Poor Law); healthy, state-subsidised housing available for people on low incomes and a uniform largely free NHS, paid for from taxation. It is an immense contribution to the health and wellbeing of millions.

Amongst his many other achievements, the modest Addison established the Medical Research Council, was instrumental in setting up professional nursing registration, and, in the 1940s, helped to sow the seeds for England's national parks (Morgan and Morgan, 1980). In the 1930s, he was a leading light in the Socialist Medical Association, which gave humanitarian and medical assistance during the Spanish Civil War and lobbied for the creation of a free, national health service. The Dawson report of 1920 had recommended the unification of hospitals into a single medical system, but was not implemented.

Legacy and implications

This writer would contend that Viscount Addison is a true hero of public health and that his most direct contributions lie in three main areas of his life: one, from 1910 to 1918, as a contributor to national insurance legislation and post–First World War reconstruction; two from 1919 to 1922, as president of the Local Government Board, then health minister (in the shape of health and insurance reform and the 'Addison' Housing and Town Planning Act) and, three, from 1937 to 1951 as a statesman, particularly as Labour's senior member of the House of Lords and an advisor to Clement Attlee and Aneurin Bevan on the setting up of the NHS and allied welfare services. One of his most important roles was to encourage Bevan, Labour's health minister, with a housing portfolio, to be far more bullish on council house building.

The periods in Addison's public health career can be summed up as:

One, Liberal reformer – 1910 to 1918
Two, disillusioned minister – 1919 to 1922
Three, elder statesman – 1937 to 1951

He was also, in his last few years, a popular dominions secretary, one of the secret masterminds of Britain's nuclear defence policy (Morgan and Morgan, 1980) – unlike many of his left-wing Labour colleagues, he was against appeasement with Hitler – and, in 1946, was the first Labour politician to be made a Knight of the Garter. When Addison died of pancreatic cancer, in 1951, there was a memorial ceremony at Westminster Abbey. Messages of condolence were received from hundreds of people from the famous (including Churchill and Beaverbrook) to the humble.

Achieving political change can be an exciting white-knuckle ride. It requires a mixture of bravery, brute force and finesse combined with skills of co-option and persuasion. When Christopher Addison entered the House of Commons for the first time as MP for working-class Shoreditch, the Liberals' People's Budget of 1910 had recently been passed, avoiding a constitutional crisis and giving Asquith's party a mandate to continue its programme of anti-poverty reforms, begun in 1906 (Hatchett, 2011).

At this time, Suffragette militancy was at its height. The Liberals' hold on power was fragile, depending on the support of Labour and Unionist MPs, although the party was predominantly pro-Irish Home Rule. High on the party's list of unfinished and needed reforms was its National Insurance Bill, mainly the brainchild of Chancellor of the Exchequer, David Lloyd George, following a fact-finding visit to Germany in 1908.

The bill's proposition was simple. All wage earners could join a compulsory contributory sickness and unemployment insurance scheme. They would donate 4d a week, their employer 3d and the state 2d (hence the strapline 'ninepence for fourpence'). Revenue from the scheme would be used to fund a network of tuberculosis sanitoria, run by local health committees. There were problems: Conservatives were ideologically opposed, unions and friendly societies ran their own schemes and did not want competition and doctors were worried about working for the state and their level of remuneration.

In breakfast meetings, Addison supplied factual memoranda for Lloyd George and he used his well-established links with the British Medical Association to help bring doctors on board. The bill was duly passed, winning the chancellor's gratitude and admiration. In reward, Addison was appointed in February 1912 to the Tuberculosis Commission, chaired by Waldorf Astor; it led to public education on the 'white plague' (the main preventable killer of the poor) and funding for an important new body which has proved of lasting value: the Medical Research Council.

Addison's success led to him being promoted to Lloyd George's junior, when the Welshman became munitions minister in May 1915. His job was to help pep up production, using plans, budgets or 'costings' and targets – innovations then, but grist to the doctor's mill and highly successful. He was still on the way up; he was made a privy councillor in June 1916 and, in July, he became munitions minister himself (somewhat ironically for a doctor). In this role, he facilitated improved and healthier working conditions for munitions workers and was involved in commissioning state-funded workers' housing estates (an experience useful to the creation of council estates). Two examples of 'munitions estate' where road layouts and amenities, as well as houses, were carefully planned were Rosyth in Scotland and Well Hall, in Greenwich, for workers at the Woolwich Arsenal (Morgan and Morgan, 1980) (see Figure 11.1).

All of this was 'war socialism' in action – unprecedented state control of working conditions, housing and the means of production. Addison's First World War experience profoundly influenced his subsequent political ideology within the Labour Party. Why wouldn't you want the state to be in control if it was more effective than the private sector? Lloyd George, Coalition prime minister from December 1916, promoted Addison to minister for reconstruction in July 1917.

Lloyd George now created a 'garden suburb' of trusted advisors (Grigg, 2002). Government re-organisation was the order of the day, run along quasi-presidential lines that were new to Britain, in preparation for a new, post-war settlement. Nobody wanted Bolshevism in Britain – the wartime labour disputes on the Clyde and in the coal rail industries had been alarming – but it didn't have to happen. Clever, forward-looking men like Addison were at a premium. There were even press departments.

Addison was now preoccupied with the pressing demands of demobilised soldiers, such as where would they live, what would happen to the economy

Figure 11.1 Well Hall/Progress estate, Eltham, London
Source: © Jill Stewart 2016.

when ramped-up war production was geared down and where would the materials and money come from for the promised reconstruction? He was promoted again, to president of the Local Government Board, in January 1919 when the war had ended. This was the all-powerful central government department that ran local government and administered the poor law workhouses, asylums and hospitals, or dispensaries (Hatchett, May, 2015).

As health administration was an integral part of this role, it was a natural place for Addison to be. Since 1916, there had been an expectation that the poor laws would be swept away after the war and the antiquated LGB replaced by a new health 'ministry' with a dedicated health minister. The LGB's last presidents, Walter Long, Hayes Fisher and Auckland Geddes, had been obscurantists, in Addison's view, defending the status quo and standing in the way of progress (Morgan and Morgan, 1980).

From the beginning of 1919 his painstaking task was planning the succession of LGB powers to the new ministry and adding new ones, such as maternity and nursing services and national insurance, through primary legislation. In practice, poor law institutions would not disappear magically when the ministry began, leaving a hotch-potch of provision that would last until after the next war.

One of the health minister's key roles in the brave new post-war world would be – a massive innovation – 'the prevention of illness' so, naturally, he would also run a national housing programme, tasked to seek out and destroy disease-incubating slums and to build the 'homes fit for heroes' that had been promised by Lloyd George in a speech in Wolverhampton (despite chronic national shortages in building materials and labour). To this end, Addison and his civil servants were working up a comprehensive and revolutionary new housing and town planning bill.

The Ministry of Health Bill received royal assent on 3 June 1919. The new health minister began his job the following month. As expected, the man of the hour, custodian of the nation's hopes, was a Lincolnshire farmer's son, the Right Honourable Christopher Addison. He was at the top of the curve. Within less than two years, the dream would turn sour. Addison would be pariah in the eyes of the press, disowned by his mentor, Lloyd George, and humiliated.

The reasons for Addison's spectacular political downfall were embedded in the technical details of the Housing and Town Planning Act of 1919, in the rapidly changing political and economic landscape that followed the Great War and in Lloyd George's cynicism, or you could call it pragmatism.

The Addison Act, unlike its predecessor, the 1890 Housing of the Working Classes Act, was designed to underwrite almost 2,000 councils building half a million houses (using small local contractors) through an exchequer subsidy (councils could levy a penny rate, but no more) or by means of local bonds, which were never widely adopted. Post-war, prices were rising rapidly. The Treasury was increasingly concerned that its subsidy, encouraging councils to build 'affordable' workman's cottages, templated on the Tudor Walters design manual, was potentially unlimited (Morgan and Morgan, 1980).

The presses, owned by the newspaper barons, Lords Northcliffe and Rothermere, were suddenly obsessed by public profligacy. The gentle and previously obscure figure of Christopher Addison was now portrayed as a spendthrift (a member of an out-of-touch and overpaid elite) and mercilessly pilloried. Rothermere even funded a political party, called the Anti-Waste League, which contested seats in by-elections. In response, Lloyd George, who the papers would turn on next (forcing his resignation in 1922, amidst allegations of corruption) hung out his friend to dry. He accused Addison in Parliament of 'an unfortunate interest in public health' and of being 'too anxious to build houses' (Morgan and Morgan, 1980).

Led by the wishes of his Conservative Coalition partners, the prime minister effectively abandoned the new housing programme. In March 1921, he transferred Addison to a far lesser job as 'minister without portfolio'. Later that year, Lloyd George appointed a committee led by Sir Eric Geddes to find massive savings across government, recommended as £175m by chancellor Sir Robert Horne, for 1922–23 (the 'Geddes axe'). Addison initially took the job, but then resigned from government. A year later, he lost his seat. He was now set on the path of socialism, campaigning for Labour from 1923.

As he explained bitterly in *The Betrayal of the Slums*, published in 1922 (Addison, 1922), the grant devoted to his housing act, which could have saved the country millions because of all the people prevented from becoming ill, had been arbitrarily restricted to £200,000, spread across the whole of Britain (see Table 11.1). Meanwhile, the country was spending more than £200m annually on war services, including in Mesopotamia (present-day Iraq), set against only £2m on the treatment services for TB. How could this be right? 'The cost of this neglect', he argued, 'is that [we are] committed to

Table 11.1 Spending on warfare versus welfare, 1922–23

	Expenditure	*Taxation per head*
Expenditure on war services including Mesopotamia and Palestine	£233,153,000	£4 18s 0.5d
*Housing (Great Britain) including £5,750,000 subsidy to private builders	£11,336,724	£0 5s 3.5d
Treatment of tuberculosis (Great Britain)	£2,124,000	£0 1s 0d
Propose grant to redemption of slums	£200,000	£0 0s 1.5d

* The subsidy to private builders will practically terminate with the present financial year.
Source: Addison (1922).

an increasing expenditure on combating the results of these deplorable housing conditions' (Addison, 1922, p. 127). And he commented, in a Parliamentary debate in February 1922: 'It is useless to give additional sums of money for the treatment of tuberculosis and at the same time cut down the money for dealing with slums.' Exactly the same rationale for housing investment can be made, and is being made, by public health and housing campaigners today.

Conclusion

So what went wrong? What can we learn from what Addison regarded as his biggest political failure – his 'betrayal of the slums'. We can see recurring patterns in his political life. Addison was a 'detail man' and a behind-the-scenes operator; he was not someone to command a public meeting or to be a front window for a radical policy. He did not have an orator's skills or a powerful prose style.

In his political lifetime, he served as lieutenant to three hugely more powerful and well-known men – Lloyd George, Ramsey MacDonald and Clement Attlee. In the first two cases, the experience led to disillusionment. His expansive idealism was defeated by realpolitik; in 1922 and 1930 his plans were shredded when savage spending cuts were imposed in the name of financial austerity. But for all that, many of his achievements did have a lasting and positive impact and we can learn from them.

In the final phase of his career, as hand-maiden to Attlee and post-1945 Labour, he was actually successful. He was not carrying the reform programme himself now; he was, instead, a wise, background advisor and his vast experience, particularly of the 'command economy' of the First World War, proved genuinely useful to the new government with its vigorous and impatient socialists, Bevan, Cripps and Dalton. When prescription charges were imposed in 1951, Addison, the pragmatist, tried, unsuccessfully, to persuade Bevan, now minister of Labour, from resigning from government in protest.

In the first 'failure', Addison's modern-sounding argument for upstream spending on building healthy homes to save money down the line was, apparently, disregarded and abandoned. But that is hardly unusual in British politics, which are dominated by short-term priorities. This all-too-familiar process was repeated recently, when the coalition government after 2010 summarily discarded its fig-leaf conversion to public health spending. The recent Wanless and Marmot reports (2002 and 2010), on the need to address growing health inequalities through primary investment, were simply forgotten.

So what do we learn from this public health champion of the late 19th and 20th centuries? Well, several generations of families certainly benefitted from the handsome and now expensive council houses in Addison Way, near Watford, before they were abandoned to the free market. The children who grew up in them were healthier and happier than they would have been if there had not been a large-scale house building programme after the First World War.

Lots of other houses were built under the Addison Act, mostly by councils but some by housing societies and associations. The final tally is reckoned to be more than 200,000 and more than a million homes for rent were built as a combined result of the housing initiatives between the wars, making up 10% of the country's housing stock. They did make a difference. So it is worth dreaming. We will need more Christopher Addisons for those times in the political cycle when self-interest and private wealth no longer dominate public policy, against all other considerations, including what was reckoned by Cicero as the highest law of politics, 'the health of the people'.

Bibliography

Addison, C. (1922) *The Betrayal of the Slums*. London: Herbert Jenkins.

Addison, C. (1933) *Socialist Policy and the Problem of the Food Supply*. London, SW1: The Socialist League.

Grigg, J. (2002) *Lloyd George: War Leader*. London: Allen Lane.

Hatchett, W. (2011, December) A National Health Service: The Reforming Measures of the "new Liberals" Dominated Political Debate at the Beginning of the Century. *Environmental Health News*, p. 36.

Hatchett, W. (2015, March) The Home Front: The Coming of the War Had Produced a Social and Economic Earthquake That Had to Be Dealt With by Emergency Measures. *Environmental Health News*, p. 36.

Hatchett, W. (2015, May) The Home Front. *Environmental Health News*, p. 40.

Ministry of Health debate. (1922, February) House of Commons, 27 February 1922 at 6 pm. Hansard online 1803–2005. http://hansard.millbanksystems.com

Morgan, K., and Morgan, J. (1980) *Portrait of a Progressive: The Political Career of Christopher, Viscount Addison*. Oxford: Clarendon Press.

Stewart, J. (2016) *Housing and Hope*. https://itunes.apple.com/gb/book/housing-and-hope/id1138338603

Recommended websites

http://spartacus-educational.com/SPsma.htm

A useful account of the Socialist Medical Association, of which Addison was a member

https://en.wikipedia.org/wiki/Christopher_Addison,_1st_Viscount_Addison

Addison's Wikipedia entry www.bodley.ox.ac.uk/dept/scwmss/wmss/online/modern/addison/addison000.html Catalogue of the papers of Christopher Addison, 1st Viscount Addison (1869–1951), compiled by Hannah Lowery. In addition to the comprehensive catalogue, contains a biographical introduction http://hansard.millbanksystems.com/people/dr-christopher-addison/

Hansard online. Addison's constituencies and offices, titles and contributions to Parliament, listed by year, from 1910 to 1951.

12 Margery Spring Rice

Throwing light on hidden misery

Deirdre Mason

Introduction

A woman who was niece to both Dr Elizabeth Garrett Anderson and Dame Millicent Fawcett was unlikely to be destined for a retiring life. A family that included a pioneering GP and a stalwart of the women's suffrage movement was likely to produce a determined character with a mission to effect change in the community and particularly, change for women.

Margery Spring Rice was born Margaret Lois Garrett in 1887 to progressive London parents who took careers for women as the natural order of things. She studied first at Bedford College and then, from 1907 to 1910, at Girton College, Cambridge, where she read moral sciences. She went on to train as a factory inspector but, as was so often the case, marriage to a young officer in 1911, Captain Charles Coursolles Jones (later killed in World War I), and caring for the children they had together interrupted her fledgling career (Dunkley, 2004).

A second marriage, to financier Dominick Spring Rice, resulted in further children. Nevertheless, Spring Rice threw herself into public life and by the 1920s, was honorary treasurer of the Women's National Liberal Federation. It was this connection that alerted her to terrible conditions of poverty and overcrowding in North Kensington and resulted in her decision to set up a birth control clinic, the North Kensington Women's Welfare Clinic, in 1924. This was one of the earliest and biggest voluntary clinics of its kind and followed on from the work of the pioneering but now rather "difficult" Marie Stopes. It also opened in the same year that the Labour Party Women's Conference had, according to socialist, early Fabian Society member and London School of Economics and Political Science co-founder Beatrice Webb, a "discussion on birth control" which according to Webb's diaries was judged a success (Debenham, 2013).

For researchers into public health today, Margery Spring Rice's approach to running this clinic, with the woman she appointed as medical officer there, Dr Helena Wright, is a gift. Because she was an internationalist and believed in co-operation, the archives at the Wellcome Library contain detailed memoranda and typescripts of surveys carried out in co-operation with an American academic, Norman E Himes MA. Thanks to this work, we can see for ourselves the problems that led Spring Rice to set up a committee to investigate the health and conditions of working-class wives in the home. For example figures from a Himes survey into nine British birth control clinics that included the North Kensington clinic showed that, out of 3,296 cases examined, the average number of pregnancies per client was four but the average number of living children resulting from the pregnancies was only 3.17. Himes's survey work also found clear and quantifiable evidence for what one might reasonably expect, that the more pregnancies a woman had, the more babies were miscarried or else lost at birth.

Something was going seriously wrong, too, for the mothers. Between 1923 and 1933, the maternal mortality rate rose by 23% (Todd, 2015). It all added up to a picture of dead babies, mothers dying in childbirth or as a direct result of it, and general ill health, but other than these strictly medical statistics, where was the wider research into what was going wrong in these women's lives? It was Spring Rice's determination to get to the bottom of that that led to the work that came to define her career. That was *Working-class Wives – their Health and Conditions*, which was Spring Rice's account of the findings and recommendations of the Women's Health Enquiry Committee.

Sources of information

Published sources are vague about the precise genesis of the Women's Health Enquiry Committee, which came into being in 1933 with Margery Spring Rice as its Honorary Secretary. Spring Rice had in print at any rate the modesty characteristic of well-bred women of that period and her own account of the committee, published in *Working-class Wives*, describes it as coming into being, with a remit, and with a list of the women on it. Nowhere there does she explain the origins of the committee, who set it up, when the elections were held and who decided on its remit. As a result, references to this committee in other books or articles state only the date it came into effect and that it was a non-political, voluntary committee.

Nor do there appear to be any surviving papers from this committee in the archives. However, pasted into the cover of Spring Rice's own copy (now in the Wellcome Library's archives) of the original Pelican Books edition of the survey's results is a hand-written note signed by M L Spring Rice and dated February 1962. This explains that the committee "was hatched and formed by Eva Hubbach and myself, after a few years' experience at the

North Kensington Women's Welfare Centre of the appalling ill-health of a very large number of the women who came for B.C. (birth control) advice." This note also outlines, very briefly, the methodology, to be described further on.

Another letter in the archives clarifies that Spring Rice was considering the need for this committee of enquiry as early as 1931. The letter, dated 5 May 1931 and written to Margery Spring Rice, came from Beatrice Webb, signed Mrs Sidney Webb, offering to meet Spring Rice in the Senior Common Room of the London School of Economics to talk through her ideas for the committee (outlined in a previous letter from Spring Rice to herself, as a hand-written note from Spring Rice in the archives makes clear).

Legacy

It took until 1933 for the committee to be set up, but the need for it had received a boost via some new and alarming official government statistics. Dr (later Dame) Janet Campbell, who was senior medical officer for maternity and child welfare at the then Ministry of Health, and also chief woman medical adviser to the then Board of Education, had overseen a Ministry of Health committee on Maternal Mortality and Morbidity which reported in 1932. As figures collated from the Registrar General Reports for England and Wales between 1850 and 1970 show, the maternal mortality rate was, for the early 1930s, between 40 and 45 maternal deaths per 1,000 births. That compares with a rate in 1860 of just over 40 deaths per 1,000 with intervening peaks and troughs (Chamberlain, 2006).

The 1932 morbidity report said clearly that difficult births and impaired health after childbirth were caused as much by neglected health during early womanhood as by lack of care during pregnancy and childbirth, and that ill-health directly resulting from childbirth and lack of pre-natal and post-natal care often continued after the patient had passed out of the care of the maternity services.

Spring Rice and her colleagues also noted the increases in claims both for sickness and disablement benefit under the National Health Insurance Acts shown in the Reports by the Government Actuary (1930 and 1932).

The Women's Health Enquiry Committee (WHEC) was finally set up in 1933 as a voluntary and non-political committee to investigate the state of health of women in Great Britain, in particular married working-class women. The trades union movement had already carried out a considerable body of useful work on the health and conditions of the female workforce, but the health of the wife who remained at home to take care of husband and children formed no part of these studies. The government had also focused on the health of the male workforce up to this, so a huge section of the populace in terms of its health and welfare had been invisible.

WHEC was chaired by Gertrude Tuckwell, CH, JP, London's first female magistrate, a campaigner for female factory inspectors and for protecting women from industrial accidents and diseases. Spring Rice was elected honorary secretary. Also on this committee was Amy Sayle, a prominent female public health officer and chair of the Women's Public Health Officers Association. Other members were drawn from bodies such as the Women's Cooperative Guild, the Women's National Liberal Federation, the Midwives Institute, the Standing Joint Committee of Industrial Women's Organisations, the National Council for Equal Citizenship and the Council of Scientific Management in the Home (later the National Council of Women). The original remit was as follows:

> The incidence and nature of general ill-health among working-class women
> Its possible causes, such as lack of medical treatment, poverty, bad housing and overwork
> How far women observe the ordinary rules of health and hygiene and the extent to which a certain amount of ill health is regarded as inevitable

In her explanation of how the committee set about gathering the evidence, Spring Rice says that the committee took care to make sure that cases from each district social or economic class were not in any way selected from women already known to be in bad health. The committee decided at first to take information from a sample of women from widely differing social conditions and occupations, different districts, married, unmarried, insured and uninsured. The information would be gathered via questionnaires, sent to women already on the lists of the organisations represented on the committee. Most of the completed questionnaires were collected by city and county health visitors, which is an indication that although the committee was not political, it had the tacit support of the Ministry of Health and local authorities. This is reinforced by the fact that when the findings were eventually published in 1939, Dame Janet Campbell wrote the foreword.

The questionnaires were detailed and covered housing conditions, number of children and pregnancy history, state of health, occupation before marriage, amount of time sent working in the home and time for leisure, and diet.

The committee had an early disappointment. Spring Rice explains that it had hoped to get a large number of questionnaires filled in by "women of a better economic and social position" and by unmarried women of all classes. These were to be the controls for the survey. However, only 60 of the questionnaires sent out to this section of the community were filled in and returned and were omitted from the analysis and eventual report.

Nevertheless, a healthy 1,250 survey returns, all from married women who did their own housework, were eventually collated and analysed for the survey. A number of reasons delayed publication, not the least of them being an altered questionnaire a year later that required the committee to start again, and the illness of the trained statistician in charge of analysing and tabulating the completed forms. By the time the survey results and Margery Spring Rice's detailed analysis and commentary finally appeared, on 7 November 1939, as a Pelican book, the great 1936 Public Health Act was in force and the country was now plunged into war.

Results

The survey is still subject to copyright law so cannot be reproduced in detail. However, it presents a picture of women at home whose needs were continually neglected in favour of husbands, children and the home. Time and again, there were cases of permanent ill health through too many pregnancies and poor or non-existent aftercare, teeth missing or in very poor condition, strained eyesight and constant headaches, chronic backache and varicose veins through being on their feet continuously, and anaemia or general lassitude through poor or inadequate food. There were women who remained remarkably cheerful and even managed to produce healthy diets despite small budgets, but all too often, when a crisis hit the family such as a sick child and medicine that had to be paid for, or the rent money was short, the food budget tended to suffer. Women, too, said that often they were too tired to eat once they had fed the family, or that they generally just ate what was left over. It was also clear that most of these women had little or no time to themselves and had no lives outside the home bar a lucky few with husbands or relatives close by who would babysit while they snatched an hour to attend a fitness class or an evening lecture.

Housing conditions varied, but generally were of an inadequate to poor standard because that was all the family could afford. Cold, damp, no running water or else a cold tap on the landing, nowhere downstairs to keep a pram and lack of drying space for clothing was too often the norm.

Diet depended not only on how much the husband allowed for housekeeping but also on the availability of affordable food close by. It did not always follow that country dwellers had diets full of fresh and varied produce. A good cheap street market such as the one mentioned by a Fulham

(London) resident that had fresh produce every day was more often to be found in cities and market towns. It also depended on how efficient a manager a woman was. Unsurprisingly, the survey found that that tended to depend on the woman's state of health and wellbeing.

Conclusions

What was it about this survey that made it stand out and makes it a continuing source for articles and books about the years leading up to the founding of the Welfare State and the NHS in the United Kingdom? There had been many studies beforehand, but these tended to take problems in isolation from one another. *Working-class Wives* brought these problems together, and the commentary that follows the bare survey results is a master of understanding and the holistic overview that characterises environmental health thinking today. The Women's Institute (Summers, 2013) was shocked in the 1940s by the poor eating habits and lack of toilet training among evacuee children and saw domestic science lessons for girls in school as the answer. Margery Spring Rice saw how difficult it was to train a toddler when the only lavatory was down several flights of stairs and at the end of a noisome yard. She also saw that with the "appalling shortage of facilities and domestic conveniences" at home, it was often difficult to cook in any meaningful way. Her proposed solution was to change the circumstances first of all, and then see what further education might achieve.

How influential was this survey? After it appeared, Margery Spring Rice was invited to speak at the Eugenics Society (now a concept distasteful to the modern reformer for reasons of its attempted social engineering and its terrible misuse in Nazi Germany but in the 1930s, popular among prominent social reformers and seen in terms of producing healthier and "brighter" people rather than advocating any destruction of those who did not meet the standard). *The Lancet* of 3 February 1940 published an abstract of her book, and *The Eugenics Review* (32/2 1940) did the same.

Of itself, this is a modest reaction but the key to the report's enduring influence lies in the recommendations. Spring Rice's report recommended an extension of the social services affecting children and a system of family allowances paid to the mother (this last recommendation was not supported by the committee's chair, Gertrude Tuckwell, as is made clear in the report). It also recommended the extension of the National Health Insurance system to cover the wives and dependent children of insured men. There were other recommendations covering housing subsidies, establishing women's clubs for recreational holidays and leisure, and agricultural subsidies to be concentrated on "protective" foodstuffs so that every working-class home could afford healthy food.

During the 1930s, Sir William Beveridge was already working on his vision for a welfare state and National Health Service. He based his eventual recommendations on social surveys that had been carried out between World Wars I and II. Given Margery Spring Rice's connection via Beatrice Webb with the Fabian Society, which submitted evidence to Beveridge, the make-up of her committee of influential women and the support of Dame Janet Campbell, it is highly unlikely that Sir William would not have been aware of *Working-class Wives*.

In the National Archives (catalogue Ref: cab/66/31/27) there is an actual typescript summary dated 25 November 1942 of the Beveridge Report. Recommendation 6 calls for the recognition of housewives as a "distinct insurance class of occupied persons with benefits adjusted to their special needs". They were to have benefits in their own right if their husbands were unemployed or disabled, and if they had jobs, they were to have "special maternity benefit in addition to grant" but lower unemployment and disability benefits.

It was the National Health Service that brought about so many of the improvements that Margery Spring Rice and her committee saw as essential to the wellbeing of working-class women in the home. For the first time, they and their children had equal status to their husbands and fathers when it came to health care. Beveridge even advised that women who needed in-patient treatment at hospital should be eligible for a home help to take care of dependent children at home. The Spring Rice report shone a spotlight on shameful conditions of poverty and inequality endured by a whole section of housewives who had been left to suffer in silence. Never again could any post-war government claim ignorance of the lives and needs of these women and their families.

Acknowledgements

I would like to thank the staff at the Wellcome Library for their help in making crucial papers available to me that clarified several points about the genesis of the Women's Health Enquiry Committee and that added detail about Margery Spring Rice's work at the North Kensington Women's Welfare Clinic.

Bibliography

Chamberlain, G. (2006) British Maternal Mortality in the 19th and early 20th Centuries. *Journal of the Royal Society of Medicine*, 99(11), 559–563.

Debenham, C. (2013) *Birth Control and the Rights of Women: Post-Suffrage Feminism in the Early Twentieth Century*. London: I. B. Taurus, pp. 132–153.

Dunkley, S. (2004) *Rice, Margaret Lois Spring (1887–1970)*. Oxford Dictionary of National Biography, Oxford University Press, 2004. www.oxforddnb.com/view/article74760 accessed 22 April 2016.

The Fawcett Society (2015) www.fawcettsociety.org.uk

Spring Rice, M. (1939) *Working-Class Wives: Their Health and Conditions*. Harmondsworth, Middlesex, England: Pelican Books. Reprinted under the same title by Virago Ltd, London in 1981, with additional material by Cecil Robertson and Barbara Wootton.

Summers, J. (2013) *Jambusters*. New York: Simon & Schuster, pp. 92–93.

Todd, S. (2015) *The People: The Rise and Fall of the Working Class 1910–2010*. London: Hodder & Stoughton.

Wellcome Library archives: Margery Spring Rice SA/FPA/SR24C/7 Box 667: Norman Himes Papers and Correspondence SA/FPA/R15/24 Box 665, and Letters of historical interest (Margery Spring Rice) 1920s-1960s SA/FPA/SR5. Wellcome Library, 185 Euston Road, London NW1 2BE.

National archives

Beveridge, Sir William (1942) Social Insurance and Allied Services: report by Sir William Beveridge. Presented to Parliament by command of His Majesty November 1942. HMSO CMND 6604. www.sochealth.co.uk/national-health-service/public-health-and-wellbeing

Blackburn, Sheila (1995) How Useful Are Feminist Theories of the Welfare State? *Women's History Review*, 4(3), 369–394. Routledge Taylor & Francis Group.

Background reading

Blythe, R. (1973) *Akenfield*. Harmondsworth, Middlesex: Penguin Books.

Forrester, H. (2011) *Twopence to Cross the Mersey*. London: HarperCollins.

Mackenzie, S. (1989) *Visible Histories: Women and Environments in a Post-War British City*. Montreal: McGill-Queen's University Press, p. 24 on conditions of housing.

Palmer, D. (1986) *Women, Health and Politics 1919–1939*. WRAP project, University of Warwick. www.wrap.warwick.ac.uk

Pascall, G. (1997) *Social Policy: A New Feminist Analysis*. Hove, UK: Psychology Press.

Smith, F. B. (1990) *The People's Health 1839–1910*. London: Weidenfeld & Nicolson Ltd.

Worth, J. (2012) *Call the Midwife*. London: Phoenix Paperback, Orion Books Ltd.

13 Berthold Lubetkin

'Nothing is too good for ordinary people'

Ellis Turner

Introduction

Berthold Lubetkin's legacy continues to impress and inspire public health initiatives in Islington, London, where his architecture for housing, health centres and wider leisure facilities continues to have a positive effect in particular on Islington's residents' outlook and wellbeing.

Born in Georgia in 1901 Berthold Lubetkin pioneered modernist design in Britain in the 1930s. His work in Islington also included the Spa Green and Priory Green Estates and Bevin Court. In the 1930s he set up the architectural practice called Tecton (an abbreviation of the Greek Architecton), designing many pioneering housing estates before his plans to replace Finsbury's slums with more modern housing designs were deferred in 1939 by the onset of World War 2.

Perhaps his overwhelming belief was to be a key part of the design of some of the best social housing that modern architecture and construction technology could provide and to bring the very best environments to whom he considered the worthiest, the ordinary person. No doubt some of these influences were nurtured whilst bearing witness to the socialist revolutions during his study in Moscow and Leningrad in 1917. Other influences to his work include traditional textile designs and processes from his Georgian ancestry. His work was also inspired by Russian constructivism, facades played a key role in the elements of his work such as chequerboard frontages, musical designs and bold use of colour. His designs were exciting and literally enlightening (Coe, 1982; Round, 2016; Wikipedia, undated).

Most of this chapter is based on living and working in the modern London Borough of Islington, visiting and passing buildings he designed and being privileged to receive "universal access to healthcare" services from the striking Finsbury Health Centre in Pine Street and a growing, snowballing interest in what else Lubetkin has designed. This chapter therefore represents an academic, but also a developing organic, personal journey of discovery including on estates such as Priory Green, Spa Green and Hallfield as well the Finsbury Health Centre.

Legacy

Today the revival of interest in and link with housing and health could not be any more explicit. So too is the role other Public Health practitioners can play to contributing to positive health outcomes and reducing health inequalities caused by poor housing. Environmental Health Practitioners' (EHPs) renewed focus in the broader determinants of health increases as the supply of good decent social housing decreases. Never before has the pioneering work of Lubetkin been so relevant and provided such impetus in today's public health agenda.

Lubetkin didn't need to prove the link between housing and health as he was acutely aware of his task of providing houses for heroes. Lubetkin arrived in Britain at a time of great social deprivation and gathered together a group of architectural students. He made it his mission to fight for better housing conditions for these heroes. He strongly believed that Tecton were best placed to attack the housing and health problem they had been set to solve, as unfit housing and disease were widespread at the time (Reading, 1982).

Perhaps Lubetkin's greatest contribution to public health is the design of some of the finest built social housing in London. The provision of well-designed, safe, warm, hazard-free and affordable housing is still as relevant to public health today as it was when Lubetkin arrived in Britain. Lubetkin set a standard not just for form but for function too; other council estates he designed in London include Highpoint apartments in Highgate and Hallfield estate in Paddington with their sweeping balconies and naturally lit vistas (Wikipedia, undated).

Lubetkin was part of the Modernist movement in London and the modernists were to be influential in designing living environments. Many were committed socialists who sought to deliver an entirely new way of living using the latest technologies with socially progressive amenities, leaving no gaps in need. They were particularly committed to enhance the lives of residents and to move away from traditional approaches. Of particular significance is Le Courbusier, who provided immense impetus in architecture and furniture, but in London in particular there were others commissioned by progressive councils and individuals, such as the Bermondsey Health Centre (under the auspices of Alfred and Ada Slater) and the Peckham Pioneers Experiment (led by Doctors George Scott Williamson and Innes Pearce), who built beautiful centres to deliver local health services.

For housing, Wells Coats designed the Isokon (Lawn Lane) flats and Embassy Court, whilst Maxwell Fry and Elizabeth Denby designed – with tenant participation, then a radical idea – Kensal House in Ladbroke Grove. These architects were in turn to later influence high-rise developments, most particularly large concrete council estates such as Park Hill at Sheffield, flats by Denys Lasdun, as well as the later brutalist architecture of Chamberlin, Powell and Bon's Barbican and Goldfinger's Trellick and Balfron Towers (see also Stewart, 2016).

Whilst part of this movement and its ideals, Lubetkin's design aesthetic includes soft-patterned, blocky, tessellated style and use of building elements, particularly glossy tiles as seen at Spa Green Estate, Finsbury Health Centre and Priory Green Estate. This style is often mimicked by other municipal architects especially in Islington and a lot of Public Houses also built in the 1930s.

However, well-designed social housing wasn't the only contribution to modern public health that Lubetkin made and one of his most significant contributions was the Finsbury Health Centre. We shall now look at some of his significant architectural achievements in the modern London Borough of Islington, which continues to pioneer contemporary approaches to public health.

In 1982 Lubetkin was awarded the RIBA Royal Gold Medal and died not long after in 1990. His work on the Hallfield and Spa Green Estate is still being remembered and showcased today as models of good practice (Wikipedia, undated; Round, 2016).

Finsbury Health Centre

Finsbury Health Centre contains, as Lubetkin stated: "an entrance hall flooded with light, through a wall of glass bricks, clean surfaces and bright colours to produce a cheerful effect" (Wikipedia, undated).

Many have benefitted from the range of health care treatments available at the Centre and the therapeutic nature of the design aesthetic itself allowing both light but also privacy. Some describe it as a spaceship that has landed from the future, and this description is as physically apt as its design and aim: to achieve noble public health goals through lofty social ambitions of universal free access to health care and to be complemented by the provision of the finest housing architects could design. Finsbury Health Centre is a futuristic iconic "spaceship" of social progress that landed from the future and whose philosophy is here to stay (Lewycka, 2016).

Commissioned by the Labour Council of the Borough of Finsbury and completed in 1938, it had revolutionary public aspirations. One of Lubetkin's greatest legacies is perhaps that he was part of this significant moment in British social history that sowed the seed of development of today's NHS (there are also parallel health centres designed around this time; for a brief description see the 'legacy' and relationship to Bermondsey and Peckham, inner urban areas sharing similar poverty and deprivation at the time).

Finsbury at the time was a densely packed slum, rife with disease, pests and deprivation. Labour politicians were determined to make Finsbury a model of social progress, not just with housing but with education, hygiene and health too. Ironically Islington still has the highest deprivation indices of one of the London boroughs. The 1938 Centre incorporated a TB clinic, a foot clinic, a dental surgery and a solarium. Lubetkin once described how history repeats itself first as tragedy and second as a farce; this charge can still be levelled at the rise and fall of social housing and the continuous need for excellent health care provision. Today stop smoking services, the Clerkenwell Medical Practice and the Michael Palin Centre for Stammering Children Services operate from the centre.

Spa Green Estate

The Spa Green Estate features "lifts, central heating, balconies, daylight from multiple directions and a spectacular roof terrace" (Wikipedia, undated). The Minister of Health Aneurin Bevan laid the foundation stone to Finsbury's Spa Green Estate in 1946, and this included the planting of a plane tree by Princess Margaret. In 1998 the building received a Grade II* listing. Residents have reported that something about the 1930s art deco style combining squares and curves attracted them to wanting to see his work up close.

Each flat gives clear unobstructed views, and his aerodynamic 'wind roof' provided a communal area for drying clothes and another new type of social space. The flats originally included fitted kitchens, slide-away breakfast counters and ironing boards, electrical and gas appliances and a stainless steel central waste-disposal system. You can understand Lubetkin's thoughts when he said of the Spa Green Estate, "we will deliberately create exhilaration."

The site had been designated for slum clearance but was then partly demolished by German bombing similar to the site for Bevin Court. Some say it is the most complete post-war realisation of a 1930s radical plan for social regeneration through modernist architecture and that Lubetkin intended this project as a manifesto for modern architecture.

Two parallel blocks create a central plaza, which also contained a nursery school and Lubetkin's design made sure that everyone had a balcony on the street side. No flat is overlooked by another and there is no hierarchy of back and front or flats with good views and those without. All these brightly lit interiors have the bedrooms on the quieter side and a spacious living area with direct access to the balcony. Quality and equality were embedded into the design, with a civic balance of socialism and philanthropy (Wikipedia, undated).

Priory Green Estate

The Priory Green Estate was stunning by his design and scale with Lubetkin features such as large communal areas and long balconies. Sadly in the 1980s this estate became notorious for drugs and prostitution. Used needles and contraceptives would often litter the grounds of his work and open drug dealing would plague it too. It was sold to the Peabody Trust in 1998, which really turned the estate around.

However, Peabody renovated the buildings, constructed a new playground, introduced landscaping and built a new entrance block and concierge. One of Islington's commissioned services (Help on Your Doorstep), an embedded team, now has an office on site. Their aim is to make a difference to health and wellbeing in the communities by empowering individuals to overcome the barriers they face and improve their lives. Aims that would make both Lubetkin and George Peabody proud.

This chapter does have a happy ending concluding with one of the finest examples of regeneration complete with an embedded task force to remove housing hazards and to improve people's lives. There are numerous additional websites listed at the end of this chapter which readers are urged to visit.

Bevin Court

Bevin Court (Figure 13.1) is perhaps the jewel in the Lubetkin Islington crown. Despite austerity measures forcing Lubetkin to remove some of the basic amenities he had planned such as balconies, a community centre or nursery school, Lubetkin decided instead to design a healthy social space. He created a stunning staircase which forms the heart of the building. The project was initially to be called Lenin Court and the building occupies the site of the 1902–03 home of Lenin, which he and his wife occupied while in exile, editing the Russian socialist newspaper *Iskra*.

However, as the Cold War intensified the scheme was renamed Bevin Court; a bust of Lenin decreasingly popular and vandalised by anti-communists was at one stage under 24-hour police guard. Legend has it that Lubetkin buried his memorial to Lenin under the central core to his staircase. Recent lottery funding has restored the stunning mural in the communal area (Wikipedia, undated). Again, readers are referred to websites for a visual journey through Bevin Court.

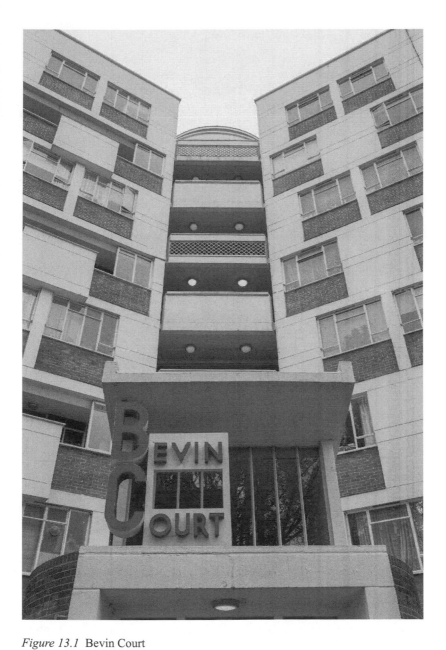

Figure 13.1 Bevin Court

Source: © Photo by Jim Gritton, 2017.

Implications for current policy and practice

As we have seen, the modernists – including Lubetkin – sought to bring radical new ideas into how we could, and perhaps should, live. With contemporary public health policy and practice, it seems timely to review some of these ideals and to understand why Lubetkin's architecture remains popular and provides desireable, aesthetic places in which to live, attend for health care and even visit in leisure time; he of course also designed the now listed Penguin Pool at London Zoo and the elephant enclosure at Whipsnade Zoo.

The way in which we use buildings and how they make us feel and behave is also important. Lubetkin surely has some influence on innovative modern health venues and healthy living centres such as the Bromley by Bow Centre, which try to engender more holistic and sustainable approaches to health (see www.bbbc.org.uk/).

The work of Lubetkin and other pioneers still serves as a reminder that our work to ensure adequate provision of well-managed and maintained housing regardless of tenure is still not done. Other key drivers prevail in today's modern public health agenda and help keep the focus clearly on housing and health.

Finally, analysis of some of the wider determinants of health on the Public Health Outcomes framework is a clear indicator of how housing is informed and informs some of the broader determinants of health. This is another example of how Lubetkin's work was cutting edge not just by design but in already achieving public health goals long before accurate health data and evidence were widely available of the impact of living in a decent home.

Conclusions

Lubetkin believed that architecture should be a potent weapon that empowers people; he certainly achieved that and by doing so helped win his battle against unfit housing and disease (Lewycka, 2016). We can still learn lessons from Lubetkin's work and his buildings serve as an (in)convenient reminder that good housing should not be beyond the reach of ordinary people and that we should continue to strive for decent housing to be a social right and not a traded commodity.

Lubetkin gave us hope, a message from the future, he taught that we all deserve to achieve and can achieve greatness no matter how ordinary we may be. His work had quality, equality and wellbeing for all embedded in his design. His healthy futuristic homes were to complement a free health service and sowed some of the seeds for a functioning NHS, and he can be considered a true modern public health pioneer.

For many, Lubetkin epitomises empowerment, hope and wellbeing, using his genius of design to be a positive and committed driving force on the side of enlightenment. Indeed, in his own words, "nothing is too good for ordinary people."

Acknowledgements

To my Aunt and Uncle Lesley and David Rock who spent many years of their lives working for the Royal Institute of British Architecture (RIBA). Their work has not only inspired me but many others both as a child and as an adult with bespoke events and conferences.

I remember sitting in a conference hall at RIBA at one of the many children's events they helped organise in the 1980s to inspire young minds hearing about exciting plans to convert the Battersea Power station site into an enormous theme park. How apt that some 30 years later the site is finally being developed instead for housing. The one remaining question is how much of this new build site, like many similar sites, will offer an affordable housing provision.

My aunt and uncle were both very lucky to meet Lubetkin, although he was a frail man when they met. They also helped me access several RIBA journals that had preserved conversations with Lubetkin including his RIBA Gold Medal acceptance speech. They have both helped me appreciate and be inspired by architecture, and Lubetkin's work is where both our professions have dovetailed into the public health agenda.

I'd also like to thank Anita Poulter for access, information and inspiration for Bevin Court.

Bibliography

Coe, P. (1982) Lubetkin and Tecton – Peter Coe speaking with Malcolm Reading at the RIBA 1982.

Lewycka, M. (2016) *The Lubetkin Legacy, Penguin Random House UK*. www.penguin.co.uk/books/176201/the-lubetkin-legacy/

Reading, M. (1982) A study of the working methods of Tecton. Speech given to RIBA on 22 June 1982.

Round, G. (2016) RIBA Commission and Exhibition (The Architecture Gallery). We live in the office, a façade style book RIBA Architecture.com. www.architecture.com/RIBA/Contactus/NewsAndPress/PressReleases/2016/WeLiveInTheOffice%E2%80%93ACommissionByGilesRound.aspx

Stewart, J. (2016) Housing and Hope: The Influence of the Interwar Years in England. https://itunes.apple.com/gb/book/housing-and-hope/id1138338603?mt=11

Wikipedia (undated) *Berthold Lubetkin*. https://en.wikipedia.org/wiki/Berthold_Lubetkin accessed 9 December 2016.

Wikipedia (undated) *Spa Green Estate*. https://en.wikipedia.org/wiki/*Spa_Green_Estate* accessed 7 March 2017.

Useful websites

https://bevincourt.wordpress.com/
www.amwell.org.uk/docs/History/Berthold%20Lubetkin%201901.pdf
www.architecture.com/Explore/Architects/BertholdLubetkin.aspx
www.locallocalhistory.co.uk/municipal-housing/bevin/index.htm
www.lovelondoncouncilhousing.com/2013/02/a-lesson-in-love-priory-green-
   estate.html
www.youtube.com/watch?v=PV5UNQfcfWU

14 Conclusions

Jill Stewart

Doing the right thing: the past and the future

This book has been something of a labour of love, but one which our writers have relished and it has led to wider learning and unexpected happenings, as putting extra effort into anything in life always does. During its writing, numerous fascinating people have got in touch and shared their stories, some to write a chapter, others just to 'touch base' and say how interested they are in public health history and what they have been doing locally. This fits with the wider burgeoning interest in history generally, items in the media, tracing our own ancestors, trying to find links and inspirations in our lives today. Wider than this there is a professional interest in the history of public and environmental health and more nuanced connections with how history affects today's understanding, evidence-based and professional practice.

Many committed (and still commit) their lives to painstaking work in trying to push for positive public health development, trying to make things better. People like this know with a profound and philosophical belief in themselves that what they are doing is right, in spite of what others may have said, or thought. For some people, their internal drive is overwhelmingly in favour of progress: it is precisely these people who are our pioneers in public health.

Our pioneers come from a range of disciplines – some scientific in their approach, some responding to what they saw around them, be it living and working environments or extreme poverty. What resonates now is that many of these were saying what we now know to be good for our health, based in evidence. Alongside Marmot today, these pioneers take their place in our history for pursuing what they knew to be right: it is where we live that most powerfully shapes our health and our quality of life and we need to act strongly to mitigate the worst excesses of the environments around us. Putting their own necks on the line, these pioneers instigated or profoundly affected legislation, sometimes at great personal cost, even ridicule, because – simply – they knew it was the right thing to do. That highlights the ethical and moral dimension to public and environmental health.

Some of our pioneers were concerned with creating the right environments in the first place, others were more reactive in their approach and concerned with mitigating the worst effects of environments on poor health through legal and policy change. Readers will see that many of these roles are delivered by the modern-day environmental health practitioner, working in evidence-based and effective partnerships with other professional colleagues and with local communities.

Following a call for expressions of interest, numerous people made contact, some of whom regularly spoke at their local history societies or were involved in archives around public health activities. Despite a desire to highlight some less famous subjects, there were some people it was necessary to include in a book of this type. Others emerged during this book's writing, about whom even less had been published. For some of our pioneers, it is almost impossible to understand what drove their ideas, attitudes and beliefs, particularly when based on the wrong assumptions in such a 'pre-scientific' age. We have referred to Chadwick as the Father of the Environmental Health profession in the introduction. He was a miasmist yet pioneered immense changes, although it seems he was not greatly accommodating of others' views. In contrast, engineer Bazalgette successfully pushed to provide a sewerage system in the face of strong doubt by others, making his achievement particularly impressive.

We should not forget that the Industrial Revolution started in England, and as the population continued to move in search of work to urban settings, the rapid rise of overcrowded living environments and poor working conditions was in place earlier than in many other countries. In a new public health arena, the likes of Snow sought to bring a scientific dimension to understanding what made us ill, whilst Simon in London and Dr Duncan and Thomas Fresh in Liverpool made real inroads in public health in the emerging metropolis and newly evolving administrations.

It is not surprising that poverty – a major health determinant – features strongly in this book and Charles Booth sought to understand its true meaning to people and just what it meant for their lives and housing. George Cadbury (and of course others, including Titus Salt and Ebenezer Howard) sought to provide new living environments with access to better working environments, and it is not of little surprise that this positively affected morbidity and mortality data. There were those who favoured the government building council housing and Addison pioneered beautiful examples of living environments for the working classes during the First World War, and a little later the modernist Lubetkin added his conviction that nothing was too good for the working classes, designing council flats with strong aesthetics to encourage wellbeing. How things change. Who could be our next housing and health pioneer?

We were keen to ensure that this book had wide coverage across various areas of public health, including those whose living environments were more mobile and excluded. Here we have featured pioneers who sought to promote greater equity for marginalised, mobile communities. George Smith of Coalville campaigned tirelessly against the appalling conditions facing canal boat residents during the Victorian era. He also risked his job highlighting the plight of child labour and poor working conditions, and on the theme of children's health, others including Margaret (and Rachel) McMillan as well as Margery Spring Rice, promoted child and maternal health and wellbeing, which of course remains a key public health priority, reinforced by the *Marmot Review*.

This book is far from being a complete compilation of public health pioneers, merely a taster of some who have particularly inspired and impressed our individual authors. This is what makes up modern public health and environmental health practices with the new policies and tools we have available to us in policy and front line practice and we will continue to do so. It is perhaps worth bearing in mind that contemporary pioneers could be included in another monograph in the future.

As a last comment, some of us may use the term, "call me old-fashioned" in jest. Yet reviewing the contents of this book one final time, it is difficult not to finally conclude that perhaps some old-fashioned views are more progressive, radical and forward-thinking than those held by many who are in a position to do something more about public health today. We sincerely hope that you have enjoyed reading this edition and feel inspired and re-energised in your evidence-based public and environmental health pursuits.

Index

Printed and bound by CPI Group (UK) Ltd, Croydon, CR0 4YY

18/10/2024

01776213-0004